Energy, Mass, and Their Conversions

A Physical and Pictorial Understanding

by

Mark Fennell

© 2018

Cover Photo by Antoine Rault at Unsplash

Preface

The Fundamental Entities of the Universe

What is Energy? What is Mass? How do we convert one into the other? These are the questions we will answer within this book.

Energy and mass are two of the most fundamental entities of the universe. All physical objects and all motions can be based on either energy or mass. And yet, scientists do not really know what these entities are. We can calculate them and measure them. We can make predictions and build technology based on them. Yet what these entities are, as physical objects, have been mysteries to science. Until now.

Learning the Secrets

In this book you will learn the answers to these great mysteries of science. You will know the physical nature of energy, and the physical nature of mass. You will know what these entities look like, and how they convert into the other.

Physical, Not Mathematical

It is very important to notice that our descriptions will be physical, not mathematical. The majority of physicists attempt to describe things with equations, which is no description at all. Equations are useful for predicting and building, but not for understanding.

In this book, you will see the true physical nature of energy and mass. You will also learn physical processes of energy converting into mass, and of mass converting into energy. These descriptions will be supplemented by numerous full color illustrations.

Therefore, in this book you will have the first true physical representations of energy, mass, and the conversions between them.

Unified Field Theory and Theory of Everything

These physical descriptions will of course lead us to the Unified Field Theory, and the Theory of Everything. In fact, this author has already written developed these theories.

Ask any respected scientist, and he will say that the two major pursuits for science are the Unified Field Theory, and the Theory of Everything. Both of these have now been solved. The first step is understanding the true nature of energy and mass, as presented in this book.

Series of Books Leading to the Ultimate Goal

Note that this Author has already developed and written the Unified Field Theory; which I prefer to call the Unified Energy Solution. Furthermore, this Author has also already developed the sought after Theory of Everything. The details have been discovered.

A short version of these major Theories has been written, yet the paper has been shared with only a few colleagues. Thus the short version of the solution exists, but only in the hands of few people.

In order to fully understand the Solutions to these Grand Theories, the public must be led gradually through the details of each discovery. Therefore the entire series of books published by this Author leads to the Ultimate Goal of presenting the Unified Theory, and the Theory of Everything. The book in your hands, on Energy and Mass, is one publication step in the teaching of these new concepts.

Excerpts and Future Books

Chapters Taken from Other Books

Each of the chapters in this book are taken from other books by the Author. Specifically, the first two chapters are taken from the book "New Model of the Atom". The later four chapters are taken from the book "Photons in Motion".

Both of these books are works in progress. While working on these other books, I realized the benefits of having a separate book just on Mass and Energy. As a teaching tool and reference, a separate book just on Energy and Mass would be valuable.

However, the reader will notice a few comments and references which make sense only with respect to their original books. Many of these are left in simply because this book is a condensed book of material provided elsewhere; it is easier to assemble the chapters and leave as is than to go through and re-edit every word.

Excerpted from book "Photons in Motion"

Note that many of chapters are taken directly from the book *Photons in Motion*. The book *Photons in Motion* was written first. In that book, the reader learns that Mass-Energy conversions play major roles in the motions of Electromagnetic Energy. Therefore the Mass-Energy chapters were written to lead the reader to understanding how the conversions result in the EM motions.

However, I soon realized that these concepts were important enough to stand on their own. Readers will find value in having physical descriptions of mass and energy. Therefore a separate book (the book in your hands) was created.

Regarding this book, note that most chapters were taken directly from the book *Photons in Motion*. Therefore you will see many references to Electromagnetic Energy. You may also find some references to topics "discussed later in this book" which you will not find in this book; because that phrase was intended for the original *Photons in Motion*.

More Details in book "New Model of the Atom"

Also note that much more complete details on the structure of energy are given in the book "New Model of the Atom". In that book we begin with the One Universal Energy, then develop into all other energies and objects. The reader will find a shortened version within this book on Energy and Mass.

Table of Contents

1. Energy Strings: Main Concepts   9

2. Universal Energy: Creating Energy Types and Mass   21

3. Magnetic Strings and Gravity Strings   37

4. Mass Energy Conversions   51

5. Applications of Mass Energy Conversions   77

6. Gravity Strings and Mass Energy Conversions   93

7. Review and Conclusion   99

Chapter 1
Energy Strings: The Fundamental Entities

Chapter taken from the book *New Model of the Atom.*

Introduction

Energy strings are the most fundamental physical entities of the universe. All particles, energy fields, gravitational energies, and internal motions of objects can be traced to these Energy Strings. Therefore, if we are to understand anything about the physical universe, then we must understand the Energy Strings.

Note that these concepts of the Energy String differ in many ways from traditional views of energy strings. Yet we know this model of energy string works, because we have successfully used it to explain everything in physical science. All particle structures and motions have been explained using these physical descriptions. This physical of model of Energy Strings can be considered accurate because it works everywhere.

Details of Energy Strings in Various Publications

The details of these Energy Strings have been fully discussed and illustrated in other publications, including the book "Introduction to Gravity Strings", and the upcoming book "Photons in Motion".

And though the details have been fully worked out, there are so many physical aspects of these energy strings and so many different applications that different books and articles are needed to focus just on certain aspects of the Energy Strings.

Therefore, in this book we will focus on only those aspects of the Energy Strings we need for the current topics. All other aspects of Energy Strings not discussed here can be found in other publications.

The Magnetic Energy String:
Energy Wavelettes and Mass Spots

We begin with our Magnetic Energy String. The Magnetic Energy String is actually composed of smaller units. It is an arrangement of Energy Wavelettes and Mass Spots.

Mass Spots will be designated by: ●

Energy Units will be designated by: ⬭⬭⬭⬭

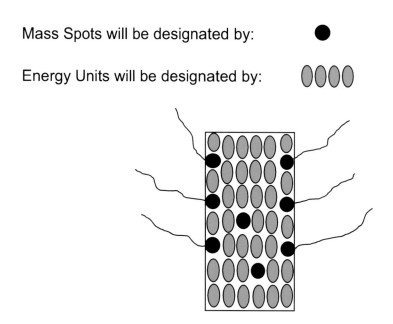

The grouping of energy wavelettes, as a whole, is the Energy String. However there are no real boundaries to an Energy string. It is really just a group of energy wavelettes moving together, as fish moving together.

There are mass spots throughout the energy string, surrounded by energy wavelettes. These mass spots were created by condensing the energy into tight structures.

In the diagram above you will also see a gravitational energy string. The gravity string is represented as a black wispy line. The gravity string is much smaller than the magnetic energy string, and only emerges from the mass spot. The inner mass spots also have their own gravity strings, but are not shown for simplicity.

All magnetic and electric field strings have some variation of this arrangement. Further details will be described below.

The Basic Composition of the Energy String

What is this Energy String? At the most basic understanding, the Energy String is a ribbon of energy. It is a stream of energy, with a beginning and an end. It is like a worm, which moves itself along through space. This is the general nature of the energy string.

The Energy String is essentially an independent flowing entity that can travel anywhere, create particles, and travel between particles. We will understand in this chapter, and throughout the book, more of what the Energy String is and how it operates.

Energy Fish are the Basis of any Energy String

In order to truly understand the physical nature of these Energy Strings, we must get a closer inspection. We must understand what the Energy String is actually composed of, and that is a group of "Energy Fish".

The energy string is not exactly the most fundamental entity of the universe. A more fundamental entity is the Energy Units, which I refer to as Energy Fish. There are these tiny units of energy, which I call "energy fish". These units are complex 3-Dimemsional shapes, intricately folded in an origami type fashion. Yet each of these energy shapes will swim forward through a type of vibration (similar to the way a fish "vibrates" and propels itself forward). Thus, we have these tiny "energy fish" which "swim" throughout the universe.

Yet most of the time these Energy Fish group together. Just as a group of fish in the ocean group together and travel together, these individual units of energy, these Energy Fish, group together and travel together.

We call a group of fish in the ocean a "school of fish". Similarly, we can call a group of Energy Fish an "Energy String". The physical structures are very similar.

Thus: any Energy String is actually composed on thousands of tiny "Energy Fish". These energy fish can travel independently, but often prefer to travel in a group. Any one grouping of Energy Fish, traveling together through space, is – when viewed at a slightly larger scale – our "Energy String".

Difference Between Energy Type is Shape of Energy Fish

Every type of energy known to physical science is therefore based on these energy fish. The only difference between one type of energy and another is the specific energy unit or energy fish. Just as we have different schools of fish, composed of different types of fish, so we have different types of Energy Strings, which are composed of different types of energy fish.

More specifically: each type of energy is a specific 3-D shape of energy. Thus: the difference between magnetic energy and electrical energy, or between magnetic energy and gravitational energy, is the specific origami shape of the energy…the specific shape of the energy "fish". It is a combination of shape, size, and vibrational energy which ultimately makes the difference between each type of energy unit. All of these factors can be included in the basic category of "specific complex shape of the energy unit".

Once the shapes are formed, then these "energy fish" can merge with others of their own type, which creates our "group"…our Energy String. At this point, the identical energy fish, as a whole, can migrate, push, pull, and flow in never ending loops.

Merging of Energy Fish into Larger Energy Strings

Energy strings can merge together to become larger energy strings. Energy strings can grow either in thickness or in length.

Any two energy strings, of the same type of energy, can easily merge together. We can understand this by remembering that the energy string is not a solid object, but is rather a school of energy fish which swim together. Therefore, each school of energy fish can easily merge together, as the individual energy fish of each group merge together.

For example: two magnetic energy strings approach each other, almost side by side. What this really means is that two groups of magnetic energy fish are approaching each other. As the two groups of energy fish get closer, the individual energy fish begin to merge. (Think of cars trying to merge into traffic – the physical process is similar).

Eventually, all magnetic energy fish, of both groups, will have merged together…resulting in one, larger, group of magnetic energy fish. Or, if we now zoom back a distance, we see that two magnetic energy strings have merged together to become one, larger, magnetic energy string. This is the basic process for energy strings to merge together.

This merging usually occurs in one of two possible directions: from the side, and from behind. If the two energy strings approach each other approximately side by side, then the merging will most likely result in a thicker energy string. If the two energy strings approach each other approximately one behind the other, then the merging will most likely result in a longer energy string.

There are of course exceptions. Gravity strings in particular rarely grow in thickness, tending only to grow in length. However, the above is a general rule for increased dimensions after merging which holds true most of the time.

Larger Energy Strings Breaking into Smaller Segments

Note also that large energy strings can break into smaller segments. The process is basically the reverse of merging.

Remember that the large energy string has numerous energy fish, generally swimming together as a group. However, under some circumstances, some of the energy fish will be induced away from the main group. This could be do a change in the direction of flow for many of the energy fish in a region, or it could be due to an external force pulling on that group of energy fish. In either case, there is a group of energy fish which no longer is acting in unison with the main group.

Thus, this new group of energy fish eventually breaks off from the main group…and begins to travel its on its own. Now we have two smaller energy strings, each of which are independent, and are likely traveling in different directions.

Migration of Energy Strings

Energy strings move on their own, through all regions of space. We can understand this by remembering that the energy string is really a group of energy fish migrating together. Just as a school of fish moves throughout the ocean, a school of energy fish will move throughout empty regions of space. If we zoom out a bit, we see this as an individual energy strings, migrating through space. At this level, the energy string appears somewhat like a tiny worm: vibrating its way forward. This is the general motion of the Energy String.

Transfer of Energy

This motion of the Energy Strings can then be seen to explain the cause of energy transfer. From my studies of energy, I have developed the following General Principle of Energy Transfer: All processes of energy transfer are simply the act of energy strings leaving one particle and entering another particle.

For example: we begin with magnetic energy strings inside a particle. This provides the internal energy, and propels the particle forward. When two particles collide, the energy strings are shaken up, and some of the magnetic energy strings exit through the holes in the surface of the particle. This is the process for a particle "losing energy".

Now these energy strings flow where they desire. Yet the second particle is very close, and therefore the magnetic energy strings enter the second particle, through the surface holes of that particle. Now the magnetic energy strings are inside the second particle. This is the process for how a particle "gains energy".

Therefore, again, the process of Energy Transfer is simply the migration of at least one energy string from a first particle into a second particle.

Internal Energy: Pushing From the Inside

All particles have some amount of internal energy. This internal energy can now be understood as the Energy Strings residing within the particle.

Furthermore, we can now easily understand how a particle has motion, and why the particle is propelled forward. The answer is from the Driver Strings.

We have this particle with many Energy Strings residing inside the particle. Taken together, this is the total Internal Energy of the particle. All of those energy strings will be flowing forward, as described above. Yet because the energy strings are contained, these energy strings encounter a wall. What happens when you push on an object? It moves forward. The same is true here. The Driver Strings within the particle keep flowing into the interior wall of the particle. All of this pushing of the interior wall will cause the particle to move forward. This is the basic process for all particle motion.

Note that the energy strings, as groups, can flow in various directions, and thus push on any wall of the particle in any possible direction. However, due to other physical processes, most the energy strings will tend to flow in one particular direction within the particle. This wall, this direction pushed by the strings, will be the over direction which the particle will travel.

Amount of Internal Energy in a Particle

We then expand these concepts to the amount of internal energy and a particle, and the resulting motions. This is particularly important for the electron, and our various orbits above the nucleus.

Above we said that motion is caused by energy strings within the interior of the particle, pushing on the interior walls, and thereby propelling the particle forward. Yet suppose we increase the number of internal energy strings pushing on that wall? The result will be more energy units pushing, and therefore the particle will be propelled faster.

Thus, for example: if we add more energy strings to a particle, there will be more internal energy strings (and thus greater internal energy). This will result in more energy strings pushing on the wall, and therefore the particle travels with more energy. The electron will travel in that direction at a much faster rate.

In brief: more internal energy strings will push on the walls with greater energy, and provide more energy to the motions of the particle.

Gravitational Energy: Pulling from the Outside

As described above, energy strings can push particles forward from the inside. Yet energy strings can also *pull* objects forward from the outside. This is particularly true for the Gravity Strings.

The same school of energy fish can either push an object or pull an object. In the case of Gravity Strings, the energy fish are physically tethered to the particle. The fish are also tethered to each other. Then, as the school of fish swim forward, they pull the particle forward as well.

The best way to visualize the Gravity Strings to think of a team of horses pulling on a carriage, or a team of dogs pulling on a sled. Each horse is tethered to each other, and then tethered to the carriage. Thus, all of their energies are combined as they move forward. This total energy will ultimately pull the carriage forward.

The same is true for the Gravity String. Each gravity unit (gravitational energy fish) is like a Gravity Horse. Each Gravity Horse is tethered to the one behind it, and ultimately to the surface of the particle itself. This situation allows all energies of the Gravity Horses to be combined. Then, as the Gravity Horses move forward, they pull together on the particle, and ultimately pull the particle forward as well. This is how the Gravity Strings create the "pull" of gravity.

More Gravitational Energy Results in Stronger Pull

Note also that more energy units will create a stronger pull. This can occur in two ways: longer gravity strings or more gravity strings per area.

The effect of longer gravity strings is easy to visualize. Think of a standard game of Tug-of-War. You know that when you add more people to the rope, their energies can be combined, and you can pull with a greater total energy. The same is true with the longer gravity strings. When we have a longer gravity string, we actually have more of these Gravity Horses in a line, all working together along the same tether. This creates a total gravitational energy that is much greater than a shorter gravity string.

The effect of multiple gravity strings per area (what I term "density of gravity strings per length") will also create a greater total gravitational pull. Imagine several men, pulling on several ropes, each attached to a large object. When you add more men with ropes to the same region of the object, they can pull with a greater energy than one man alone. Consequently, these multiple men can pull on the object in that desired direction at a faster rate.

The same is true for the gravity strings. When we have more gravity strings per area of the object's surface, each pulling in the same direction, then these gravity strings, together, will pull the object in that direction in a faster rate.

These aspects of pulling particles with different amounts of gravitational energy will become important as we get into the details of the electrons orbiting at different regions above the nucleus.

Types of Energy Strings

Overview

There are four main types of Energy Strings: Magnetic, Electric, Gravity, and Nuclear.

Regarding electron orbits, the Gravity String is the main one to focus on. The Magnetic and Electric are important in their own ways however: as the structure of the particle, and the internal energy which propels the particle forward.

Similarities and Differences

Each type of Energy String is essentially the same type of physical entity: a group of energy units ("energy fish") which generally move forward together.

The difference is the type of energy unit (the type of "energy fish"). Thus: a magnetic energy string is composed of magnetic energy fish; an electric energy string is composed of electric energy fish; and a gravity string is composed of gravitational energy fish.

Uses of Energy Strings

Overview of The Uses of Energy Strings

These "Energy Strings", as described above, are the sources of almost everything we know of in physical science. These Energy Strings provide us with everything we know regarding energy, particles, fields, and motion.

Above we talked about the Types of Energy, and the Types of Energy Strings: Nuclear, Magnetic, Electric, and Gravitational. Now we are discussing the Uses of these Energy Strings.

There are several possible uses of Energy Strings. The most main categories of Uses of Energy Strings are:
1. Particle structure
2. Driver Strings (Internal Energy, Propelling Forward)
3. Energy Fields
4. Energy Transfer

Note that, theoretically, any of the types of Energy Strings (Magnetic, Electric, Gravitational) can be used for any of these purposes. However, in practical terms each type of Energy String is better suited to some uses than for other uses.

1. Particle structure

All particles are formed by Energy Strings, particularly Magnetic and Electric Energy Strings. The longest of the energy strings intertwine and loop into themselves. This creates an interwoven sphere of flowing energies. This sphere of interwoven energy strings is the basic structure of all particles.

The primary difference between particles is the diameter of the particle, and the tightness of the interwoven strings. (More details on particle structures will be discussed throughout this book).

2. Driver Strings (Internal Energy, Propelling Forward)

All motion of particles, and hence all motion of any object, is the result of Energy Strings inside the particles which propel the particles forward.

Every particle is hollow, due to the way the particles are formed, and therefore can contain free flowing energy strings inside the particle. The total energy of these internal energy strings is the "Internal Energy" of the particle.

Most of these energy strings push on the interior walls of the particle, which then cause the particle to move forward. Any such internal energy string which pushes on the interior wall is called the "Driver String", because such strings "Drive" the motion of the particle. Additional details will be presented throughout this book, as well as other books.

Magnetic and Electric Energy Strings are the most common types of energy as Internal Energy/Driver Strings.

3. Energy Fields

An "Energy Field" is simply any Energy String which extends outward from the particle. These strings are associated with the particle, and move with the particle, yet extend into the space beyond the particle, where they can interact with other energy strings of the same type. Note that energy field detectors work by the energy field strings of the detector interacting with the energy field strings of the particles).

The most common types of Energy Fields are: Electric, Magnetic, and Gravitational. The physical attachments of fields to particles are discussed in other books, including *New Model of the Atom* and *Photons in Motion*.

Briefly: The Electric Field String extends from the interior of the particle, extends through the holes, and out into the space beyond. This field string is physically attached to the surface of the particle, and therefore moves with the particle.

The Magnetic Field String loops through the particle – interior and exterior – something like a key chain. This Magnetic Field Loop then travels along with the particle as the particle moves forward.

The Gravitational Field Strings are embedded inside the Mass Spot. The bottom of the Gravity String is firmly embedded within the center of the mass spot sphere, then extends outward into the world. The mass spots themselves are condensed energy and are carried with the energy wavelettes of the magnetic or electric strings. (Yes this means electrical and magnetic fields have gravity strings).

Additional details on these Energy Fields will be discussed in future books.

4. Energy Transfer

All events involving the transfer of energy are a result of one simple process: energy strings leave the interior of one particle, and enter into the interior of the other particle.

Remember that all motion of particles is caused by the internal energy strings pushing on the interior walls of the particles. Whenever there is a collision between particles, some of the internal energy strings of one particle will be shaken up and tossed out through the holes in the surface. This is "Energy Loss". Conversely, when the free energy strings wander in space, they will enter they nearest particle – which is usually the one involved in the collision. Therefore, this becomes "Energy Gain".

This basic process is the root of all the experiences of Energy Transfer.

5. Energy Transfer Using Photons

Note: the process of photon emission and absorption as a method of Energy Transfer is similar to the collision of particles, yet also different. With the photons, energy strings are emitted, and energy strings are later absorbed. In this respect the process is the same. However, the particular process has many differences. These concepts are explored in other books. (The details have been developed; the books themselves are in progress).

Chapter 2
Universal Energy

Chapter taken from the book New Model of the Atom

Introduction:
The Universal Energy can Create All Things

If we want to understand the types of energy, and how each type of energy relates to each other, we must go back to the original source: the "Universal Energy".

Based on my research, I have concluded that there is a Universal Energy, a primary energy from which all other types of energy is created. And I am not alone in this conclusion. There are others who have come to the same conclusion, although we differ in the specific formulations.

I also have a theory on the nature of this Universal Energy. However, I am not ready to state this publically. For the moment, let us just state that there is a "Universal Energy". This Universal Energy is the basic energy, the primary energy type, from which all other types of energy in the universe is created.

Universal Energy Creates All the Known Energies

This "Universal Energy" can be manipulated in various ways, to create each of the known energies. This includes the main four types: Nuclear, Magnetic, Electric, and Gravitational. The basic process involves using various amounts of the Universal Energy, and manipulating the Universal Energy into different geometric shapes. This creates the specific Energy Types that we know of in our physical sciences.

This Universal Energy can also be structured and concentrated into "mass". Mass is simply concentrated energy, an entity which I call "mass spots". The process of creating the mass spot is similar to creating the other types of energy: we begin with a vast amount of the Universal Energy, then structure it in a particular way, allowing the energy to be fully compacted into one small region…the "mass spot".

Universal Energy Also Creates the Unknown Energies

In addition, and I firmly believe this, there are other types of energies which are unknown to us at this time. These energies are just as physical as the main energies we know of in science. In fact, many living organisms can detect these other types of energies (including humans). As living beings, we are designed to detect these energies. Some of us can use these energies in our material world.

Yet, as scientists, we have not been able to understand, isolate, and detect these specific energies. In time, I am confident we will learn of such physical energies, and be able to measure them and work with them in a scientific fashion.

These other energies, currently unknown yet just as physical as the known energies – are also created by the Universal Energy. These energies are created in exactly the same way as all the energy types currently known to science.

Creating Each Energy Type from the Universal Energy

We can create each of the Energy Types from the Universal Energy. We begin with a quantity of the Universal Energy. This amount of Universal Energy is folded and manipulated into highly complex multi-dimensional structures. Each one of these complex structures of Universal Energy becomes one of our Energy Wavelettes; which we also refer to as "Energy Fish".

The process is similar to Origami. In the process of Origami we take a piece of paper, and fold it into complex shapes. The process is very similar for our Energy. We begin each process with a section of Universal Energy, then we fold this Universal Energy into complex structures.

Thus, every Type of Energy begins with the same material: the Universal Energy. Yet the specific size and shape of the final energy structure will determine the specific type of Energy.

The smallest amount of Universal Energy becomes the structure of Gravitational Energy Wavelettes. The largest amount of Universal Energy becomes the structure of Nuclear Energy Wavelettes. In the middle of sizes are the Electric and Magnetic Wavelettes.

Specific Variations of One Type of Energy vs. Another

When looking at the specific variations of the Energy Fish Units, from one Type of Energy versus another, there are three main factors. All factors are inter-related. This factors are:
1. Amount of Universal Energy per Energy Fish
2. Dimensions of the Energy Fish
3. Specific Complex Structural Shape
4. Swim Rate

1. Amount of Universal Energy per Energy Fish

Every "type" of Energy (magnetic, electrical, gravitational) begins with the same basic material: the Universal Energy. However, each differs in the *amount* of Universal Energy.

You can understand this by returning to our Origami analogy. When we begin with a larger piece of paper, we can create larger origami items. We can also create more complex shapes. The same is true of our Types of Energy. With more of the Universal Energy, we can create larger Energy Fish, as well as create a more complex structure for our Energy Fish.

Most importantly, having more of the Universal Energy per energy fish, will results in that energy fish having more "energy". This may sound like a redundant statement at first, but the concept is important. For example, the Magnetic Energy fish will have lots of Universal Energy per fish. This means it will be a larger "fish", it will be a complex structure, and most important: it will have significant power when involved in any process involving magnetic energy units.

Conversely, the gravitational energy is made of smaller amounts of the Universal Energy, and therefore this Gravity Fish will be smaller dimensions, it will be a simpler shape, and it will be much weaker in any process involving those energies.

Therefore the *amount* of Universal Energy is the first and foremost factor in determining the Type of Energy. All practical effects are derived from that.

2. Dimensions of the Energy Fish

Each type of Energy (magnetic, electric, gravitational) comes in different sizes. Just as there are different species of ocean fish, which exist in different sizes, there are different types of energy fish, each of which exists in different sizes.

The main reason for this is the amount of Universal Energy which is used in the creation of each type of energy. When we begin with a larger amount of the Universal Energy then we create a larger energy fish. If we begin with a smaller amount of the Universal Energy, then we will have smaller energy fish.

Therefore, the types of energy which are composed of the most Universal Energy will have the "most energy" per energy fish, as well as having the largest dimensions.

In order of size, the Energy Fish from largest dimensions to smallest dimensions are as follows: Nuclear; Magnetic; Electric; and Gravitational. Of course, this is also the order of most Universal Energy per fish to least Universal Energy per fish.

Thus, the Nuclear Energy Fish have the largest dimensions, and have the most Universal Energy per fish. The Gravitational Energy Fish have the smallest dimensions, and the least Universal Energy per fish.

3. Specific Complex Structural Shape

Each of the Energy Types begin with the Universal Energy, then this Universal Energy is folded into complex shapes. Therefore, the Energy Types differ not only by amount of Universal Energy and overall dimensions, but differ with the specific shape as well.

We again return to our Origami Analogy. Any one piece of paper can be folded into a variety of shapes, and the same is true for our Energy: the Universal Energy can be folded into a variety of shapes.

Also note that the complexity of the structure is correlated with the amount of Universal Energy. Consider the paper used in origami: using a larger sheet of paper we can create more complex origami shapes. Similarly, using more of the Universal Energy we can create more complex energy shapes.

Therefore we will naturally see the most complex energy structures in the Nuclear and Magnetic energy fish. We will see the simplest energy structure in the Gravitational energy fish.

3b. Number of Dimensions of the Structure

The Universal Energy is folded and structured into 3-Dimensional shapes. These are complex shapes, yet 3-Dimentional. All observations in physics can be explained with 3-Dimensional geometric structures of the Universal Energy.

However, we must also consider the option that the Universal Energy can be shaped into 4-Dimensions and 5-Dimensions as well. It is very likely that some of the types of energy are indeed 4-D and 5-D structures. These variations may be discovered in the future.

4. Swim Rate

The final main difference between Type of Energy is what I call "the swim rate". Every complex shape of universal energy really is like a fish. This entity has the energy which moves itself forward.

Think of it as an origami shape that we made, then when we let it go, it moves forward on its own. This is essentially what happens with every energy fish, of each energy type. Each energy fish, every unit, is this complex structure of universal energy. Once created, this origami structure propels itself forward. Indeed, it is much like a fish, which moves forward through space on its own.

This brings us to the "swim rate". Each type of energy is the same physical structure of universal energy, propelling itself forward at a particular rate. This can be compared to a particular species of fish, propelling itself forward thorough the water, at the same rate as any other of the same species. This is the "swim rate".

More specifically: the "swim rate" for any type of energy fish, is the rate at which that type of energy fish propels itself forward.

This concept is simple enough. Therefore: all magnetic energy fish will swim at a particular rate. Then together, this group of energy fish will travel through space at that rate. On a larger scale, this will mean that each magnetic energy string will travel through space at that rate.

We should also notice that each type of energy has a different swim rate. This swim rate is based on two factors: the amount of universal energy in the energy fish, and the structural shape of the energy fish. These two factors, together, produce the specific swim rate for each type of energy.

For example: the magnetic energy fish will have a much faster swim rate than the gravitational energy fish. This difference is due to the amount of Universal Energy in each type, as well as their structures.

The swim rate may not seem important at this time. However, in more advanced studies of energy, the importance of the specific swim rates will become apparent.

Different Sizes of Each Type of Energy Fish

Overview

As discussed above, each type of Energy Fish comes in different sizes. A difference in "size" means both the amount of universal energy, and the geometric structure.

We must elaborate on these concepts, because these concepts are very important to understanding the groupings of these Energy Fish (which we now call the "Energy Strings")

More specifically: the different sizes of the Energy Fish (in both amount of Energy and in Structure), when aggregated in groups (becoming the Energy String) will produce different practical results. Therefore, we will review the concepts of Different Sizes of Energy Fish, and begin looking at some of the Practical Effects.

Energy Fish Come in Different Sizes

As stated earlier, the physical structure of each type of energy strings is essentially the same: a group of individual energy units ("energy fish") which generally travel together in a group across space.

We should note that these energy fish come in different sizes. You can think of each Energy Type as a Species of Fish. Just as there are different species of ocean fish of different sizes, there are also *energy fish* of different sizes. Each Nuclear Energy Fish is very large, like a tuna. Yet each Gravitational Energy Fish is very small, like a minnow.

In order of size, the energy fish types from largest to smallest, are as follows: Nuclear; Magnetic; Electric; Gravitational.

This also means that the each type of energy will have varying amounts of the "Universal Energy" per fish. Thus: the Magnetic and Nuclear Energy Fish are large, with significant amounts of Universal Energy; while the Gravitational Energy Fish are very small, with very low amounts of Universal Energy.

Please note that we are talking about the dimensions of the individual Energy Fish only, not the energy string (which is a grouping of energy fish).

Sizes of Energy Strings vs Sizes of Energy Fish

Remember that the Energy String is actually a grouping of Energy Fish Wavelettes. Therefore there is a difference in size options for the Energy Fish vs the Energy String.

The Energy Fish Types will exist only in a few sizes and structures. This is what differentiates the energy types: gravitational, electrical, magnetic, and nuclear.

However, as Energy Strings, all of the types of energy can exist as Energy Strings in any size. The size of an Energy String (vs Energy Fish) depends on the number of energy fish grouped together, and the size of that type of energy fish.

Thus a Magnetic Energy String can exist within an electron, or surrounding an entire planet. A Gravity String can link two molecules together, or two galaxies. It primarily depends on the number of individual energy wavelettes grouped within that string.

Practical Effects of Energy Fish Dimensions

These sizes of each Energy Fish will have practical effects. For example, given the same number of Energy Fish in each type of Energy String, the Nuclear Energy String will be extremely strong, whereas the Gravity String, of the same dimensions, will be much weaker.

Another example: when two magnetic strings approach each other in opposite directions, they will repel each other strongly. We are very familiar with this when using any magnetic blocks. This is because the individual energy units are very large, with large amounts of Universal Energy per energy fish.

On the other end of the spectrum, we have gravity strings which are made of very small (low energy) units of Energy Fish. When two gravity strings approach each other, they do not repel. Instead, one gravity string easily slides along the other gravity string, though going in the opposite direction. The reason is due to the low universal energy of each gravitational energy unit. There isn't enough energy in those energy wavelette to repel gravity fish coming the opposite way.

We will see some of these practical effects in action throughout this book, as well as in other publications in this series. In addition, there are other practical effects which will be explored in future books.

Morphing One Type of Energy into Another

Introduction: Converting Energy Types

Energy types can also be morphed into another. For example, Magnetic Energy can be converted into Electrical Energy, and Electrical Energy can be converted into Magnetic Energy.

More specifically, the complex physical shape that defines the "magnetic energy fish" can be converted into the complex physical shape that defines the "electrical energy fish". The process is essentially being refolded into a new structure.

Origami Analogy (Larger Energy Converted into Smaller Energy)

You can visualize this by again returning to our Origami analogy. You have a set of folded shapes sitting on the table. You take one of those items, and unfold it. Now you tear the paper into smaller pieces, and begin folding those. Smaller size sheets of paper...and you also choose to fold into a different shape. Thus: you have converted one origami item into another.

The same general process is true for converting one type of Energy Fish into another. If we begin with a larger energy fish, such as Magnetic Energy, we can take one of those, unfold it, and get the original Universal Energy. Now we divide that energy into smaller pieces. And refold those pieces into a set of different shapes. When we are done, this creates the Electrical Energy, or the Gravitational Energy. Therefore: we have converted one larger type of energy into a few smaller types of energy.

Origami Analogy (Smaller Energies Converted into Larger Energy)

We can also convert the other direction, from smaller to larger. We simply combine the smaller energy units into larger ones.

Returning to our origami analogy: suppose we begin with many small origami pieces, all nicely folded into their shapes. We unfold each. Then we tape those unfolded pieces together. Now we have one larger piece of paper to work with. At this point, we can create a more complex shape (because we have more paper to work with). Thus, we have created a new, larger origami piece made from several smaller ones.

The process would be similar for the conversion of smaller energy fish, such as Gravitational or Electric, into larger energy fish such as Magnetic. We begin by taking a few of the smaller energy fish from the main energy string. Then we unfold those energy fish, each one, to reveal the original amount of Universal Energy.

At this point we combine all of these units of Universal Energy close together. We merge these wavelettes together in a new way. Now we have one, larger section of Universal Energy to work with.

We take this larger amount of Universal Energy, and manipulate into a new complex shape. Thus, we can convert Gravitational Energy into Electrical Energy, and convert Electrical Energy into Magnetic Energy.

Using these methods, any one type of Energy can be converted into any other type of Energy.

Nuclear Energy Created from Magnetic Energy

One example is nuclear energy. Nuclear energy has now been found to be a form of magnetic energy. Nuclear energy does not exist by itself. Rather, the nuclear binding that occurs between protons in the nucleus exists only due to the extreme pressure. The force of the protons pressing into each other converts magnetic energy into nuclear energy, which is a stronger type of energy. Therefore we know this conversion process exists in the creation of the nucleus.

How Often Does Conversion of Energies Occur

Having shown how each type of Energy can be converted into another, we can then ask: how easy is this to do? We can also ask: how often will this occur naturally? The answers to these questions are yet to be determined. However, it is likely that these conversions of energy types exist more often than we realize.

Mass Spots within Energy Strings:
Created by the Universal Energy

Mass Spots

The mass of any object exists in the form of "Mass Spots". These mass spots are the locations from which gravity strings emerge. Also, the mass spots, collectively, make the total mass of any particle.

One way to visualize the mass spot is to think of a spot of oil in the ocean. You know that any oil in the ocean will tend to stay together, as one entity. It can also float along the ocean, as the current carries it forward. The mass spot is very similar. This spot is an entity unto itself, this tiny patch of concentrated energy.

Yet, at the same time, this mass spot exists among the energy fish which make the "string". As the energy fish move forward, they push the mass spots. Thus the mass spots travel along wherever the energy fish decide to go.

These are the mass spots. They exist in all energy strings, in all energy fields, and in all particles.

Mass Spots Co-Exist with Energy Fish in the Energy Strings

One of the important aspect of the Energy Strings is that these Energy Strings also contains mass. In other words, these Energy Strings *contain both* energy and mass.

This is in contrast to other models which say mass only exists when energy strings are crumpled.

These Mass Spots exist among the energy fish of the Energy String. Thus, wherever the group of energy fish go, so too will the Mass Spots. They all travel together.

Universal Energy Also Creates Mass Spots

The Universal Energy which creates each type of energy is also responsible for creating the Mass Spots throughout the Universe.

The basic process is the same for creating mass as creating the types of energy. We begin with an amount of the Universal Energy – in this case, an enormous amount. This vast amount of energy is structured specifically to allow energy to be tightly compacted.

Thus: this vast amount of Universal Energy is structured in a unique way, then highly condensed. This becomes the "mass spot", which is the physical entity of mass in the universe.

Creation of the Mass Spots from Universal Energy

To create a mass spot, we begin with the Universal Energy. This universal energy is restructured, and highly concentrated. The result is an extremely densely packed blob of energy…which is our "Mass Spot".

There are two ways for this to occur: 1) from each type of energy wavelettes (such as magnetic energy) or 2) from the Universal Energy directly. Both will ultimately rely on the Universal Energy being compacted.

In the first case, we take a group of energy wavelettes, such as Magnetic Energy Wavelettes. These are of course created from the Universal Energy as discussed above. Then, many of these Magnetic Energy Wavelettes are pressed together. The wavelettes then convert to a dense sphere of complex structure, which becomes a mass spot.

In the second case, the Universal Energy is compacted directly into the Mass Spot; without the creation of the Energy Units first. Although, this will also require enough external pressure to take place.

Notice the Mass Spot is usually created from the other type of energies, which is ultimately based on the Universal Energy. Alternately, when mass spots are created directly, they are created directly from the Universal Energy; this is similar to each of the other types of energies.

Therefore, whenever the Universal Energy is structured and shaped into the specific types of energies we are familiar with, the Universal Energy can also be structured and compacted into the Mass Spots.

It is for this reason that the Mass Spots are almost always found with the other types of energies. For example, in any group of Magnetic Energy Units, there will be several Mass Spots.

Mass Spot as the Highest Form of Energy Type

Notice that because the Mass Spot is ultimately made of Universal Energy (whether directly or via energy wavelettes), this mass could be considered another "type" of Energy.

As with the other types of energy (such as Magnetic), the mass spot has these characteristics:

1. Is ultimately made of Universal Energy
2. Has a given amount of Universal Energy
3. Is a complex geometrical shape
4. Has likely been morphed from other Energy Types

Therefore, we could consider the Mass Spot as a type of Energy, and being the strongest "type" of energy.

Of course there are other differences between the mass spot and the other types of energy, which makes this comparison imperfect. Yet we point this out so that you understand the progressions and related structures of the physical universe.

Percentage of Mass and Amount of Mass Each Type of Energy

The number of these mass spots per area will vary among the types of energies. This is related to the general size of each type of Energy Unit.

Therefore, the number of mass spots per area, from greatest percentage to least percentage, for each type of energy string is as follows: Nuclear (most mass spots per string); Magnetic (many mass spots per string); Electric (decent amount, but not overwhelmingly noticeable); and Gravity (almost no mass spots).

The average number of mass spots in a magnetic string, as per area, is approximately 20% mass spots. Thus, for a given magnetic string, there will be approximately 80% Magnetic Energy Wavelettes, and 20% Mass Spots, per Area. Of course this is an average value, the actual percentage can be higher or lower.

Growth of Mass Spots

Mass spots can grow in size after being created. There are two ways this can happen: 1) two mass spots merge together, or 2) individual energy fish become part of the mass spot.

The first method for mass spots to grow is for two or more mass spots to merge together. If the conditions are right, mass spots can migrate, and hit each other. At this point, they will merge and become a larger mass spot. This process can continue as long as the conditions are right.

The other method for mass spots to grow is when individual energy fish join the existing mass spot. This is a slow and gradual process. Most of the time, the energy fish swim on their own, just pressing on the mass spot without any effect. Yet in some cases, the energy fish can change structure (unfold into original Universal Energy), and then become part of the mass spot energy structure. This process becomes important when we talk about energy-to-mass conversions.

Note that if the aggregated mass spots are of significant size, then this may produce some of the curiosities noticed in cosmology. This includes dark matter and the pre-cursor to the big bang.

Conclusion and Review
for Universal Energy as Basis for All Things

Universal Energy will Create All Things

All things begin with the Universal Energy. This Universal Energy can be aggregated into different amounts, and folded into a variety of complex shapes. The amount of Universal Energy and the structure will determine the type of energy (magnetic etc).

Any one specific structure of Universal Energy can be called an "energy wavelette", "energy unit" or "energy fish". These individual energy units, these energy fish, will swim together as a group, thereby creating a single "Energy String".

Energy Type Determined by Amount of Universal Energy

Each "Energy Type" (nuclear, magnetic, electric, gravitational) is determined primarily by: 1) the amount of Universal Energy per energy fish, and 2) the specific shape of the complex structure.

Amount of Universal Energy per Fish and Practical Results

The amount of Universal Energy per energy fish is the main factor which determines one type of energy from another. Similarly, the amount of universal energy is what creates the "strength" of each energy type.

For example, nuclear energy is "strong" because the individual nuclear energy units have large amounts of universal energy. Conversely, the gravitational energy is "weak" because the individual gravitational energy units have small amounts of universal energy.

Of course the total number of energy units in the string also is a factor, yet the base factor is the amount of universal energy per energy fish.

Amount of Universal Energy Resulting in Other Factors

This amount of Universal Energy per fish will also affect the structural shape, the dimensions, and the swim rate.

Greater amount of Universal Energy will produce an energy fish which: is stronger, larger, has a more complex shape, and swims faster. Conversely, less amount of Universal Energy will produce an energy fish which is weaker, smaller, simpler shape, and swims slower.

Amount of Universal Energy and Repulsion of Opposing Energies

The amount of Universal Energy per energy fish will also determine the amount of repulsive energy when two energy strings encounter each other flowing opposite directions.

For example, the Magnetic and Electric Energy Strings will repel each other strongly. This is because of the amount of Universal Energy per energy fish in each of those types of energy. The opposing wavelettes will force the others to divert from the original trajectories.

Yet the Gravity Strings will not repel each other, because the amount of energy per energy fish is very small. The wavelettes of the gravitational energy will simply slide along each other, going in opposite directions but not repelling.

Energy Type Characteristics

Given the concepts above, each type of Energy has its own set of characteristics:

1. Nuclear Energy Fish: The nuclear energy wavelettes are large energy fish; with larger amounts of universal energy; more complex structural shape; fast swim rate; and is generally more powerful than other energy types.

2. Magnetic Energy Fish: The magnetic energy wavelettes are also very large energy fish (slightly smaller than nuclear); with large amounts of universal energy (slightly less than nuclear); a complex structural shape; a fast swim rate; and is a powerful type of energy unit.

3. Electrical Energy Fish: The electrical energy wavelette is a significant size energy fish (though smaller than Magnetic); with a significant amount of universal energy (though slightly less than Magnetic); a moderately complex structural shape; a fast swim rate; and is an energy unit of significant power.

4. Gravitational energy fish: The gravitational energy wavelette is a very small energy fish – much smaller than any other wavelette type; there is very little amount of universal energy per energy fish; it is very simple structural shape; has a very slow swim rate; and is an overall very weak type of energy.

Converting One Type of Energy into Another

Any one of the energy types can be converted into another energy type. There is of course a size gradient, and the one energy type can be converted up or down to the energy type adjacent to it on the scale.

For example: a magnetic energy wavelette can be converted into several electrical energy wavelettes. An electrical energy wavelette can be converted into several gravitational energy units.

Conversely, we can build the larger energies. Many gravitational energy wavelettes can be converted to an electrical energy wavelette; many electrical energy wavelettes can be converted into a magnetic energy wavelette; and many magnetic energy wavelettes can be converted into a nuclear energy wavelette.

The different amounts of Universal Energy, and restructured into a different shape, will have effectively converted one type of energy into another.

Energy Strings as Groupings of Energy Fish

When working with the new understanding of physics, we work mostly with Energy Strings rather than the individual Energy Fish. Yet it is the grouping of Energy Fish which composes each type of Energy String.

Each type of Energy Fish swims independently, yet prefers to swim with other energy fish of the same type. This creates a grouping of energy fish, very similar to a school of fish in the ocean. Thus, we have various groupings of each energy fish. We have groupings of Magnetic energy fish, groupings of Electric Energy Fish, and groupings of Gravitational Energy Fish.

When we zoom out for a larger scale view, each grouping of energy fish appear to us as a single "Energy String". The energy seems to "flow" through this "string", and the string as a whole propels itself forward. Thus the "string" is really a grouping of individual energy wavelettes of the same energy type.

From Universal Energy to Energy Strings of Each Type

We can therefore understand how each of the Energy Strings are formed.

1. The Universal Energy becomes each Energy Type. These are individual wavelettes or energy fish of different sizes and structures.

2. The Energy Fish of the same type then swim as a group.

3. Each grouping of one type of Energy Fish becomes one Type of Energy String.

The result is numerous Energy Strings: Nuclear Energy Strings; Magnetic Energy Strings; Electrical Energy Strings; and Gravitational Energy Strings.

Each of these Energy Strings will then be used to create all particles, all energy fields, and all aspects of motion in the universe.

Creating Mass from Universal Energy

Mass is also created from the Universal Energy, in a similar way in which each of the Energy Types are created.

Mass is simply a highly concentrated form of energy, which we can call a "mass spot". To create the mass spot, we begin with an enormous amount of Universal Energy. This amount of energy is then specifically structured to become as dense as possible, and all the Universal Energy is compacted into a small region. This compacted form of energy becomes our mass spot.

Mass Co-Existing with Energy Strings

One of the more significant aspects of this model of energy strings is that the mass co-exists with the energy.

When we look closely at any energy string, will see numerous energy fish, with some mass spots contained in various regions. The energy fish surround the mass spots and push the mass spots forward. Therefore, where the energy fish go, the mass spots will go as well.

The co-existence of mass spots with the energy fish is very important, and will produce numerous practical results. (These will be explained and illustrated in this series of books).

Percentage Energy versus Percentage Mass

The amount of energy versus mass in a magnetic energy string is approximately 80% Energy Wavelettes and 20% Mass Spots, per area of Energy String. These numbers can of course vary.

Furthermore, the average percentages in different types of energy strings will differ depending on the type of energy. For example, in the Nuclear Energy Strings the percentage Mass is much higher, whereas in the Gravitational Energy Strings the percentage of Mass is much lower. (In gravitational energy strings, there is usually zero mass).

What is the Universal Energy?

This Universal Energy is the basis for all things. Every type of known energy, every type of particle, and every type of motion can be developed from this Universal Energy. This will be demonstrated throughout the series of books, culminating in the final book "The Theory of Everything".

What is this Universal Energy? Note that I do have theories about the nature of this Universal Energy. However I will not present those ideas until a future date.

For the moment, we will accept that this Universal Energy exists. We will then explain all physical phenomenon using Energy Wavelettes, Mass Spots, and Energy Strings; all of which are developed from this Universal Energy.

Chapter 3
Magnetic Strings and Gravity Strings

Chapter taken from the book *Photons in Motion*.

Introduction

Overview
In this chapter we will be discussing the flow and merging of magnetic strings. Note that gravitational energy plays a role in these discussions.

Flow Direction
The flow direction of an Energy String is the overall direction in which the majority of energy wavelettes are traveling. This is similar to the overall direction which ocean fish are moving.

Note that the group of a fish, as a whole, can move gently left or right. These wavelettes are not traveling in a fixed direction in the same way as the photon; the wavelettes can move left or right as they travel. Yet within the energy string, within the group, all of the energy wavelettes are traveling the same direction.

Merging of Magnetic String
Energy Strings will merge only when their wavelettes are aligned in the same direction. This alignment can be either side to side, or top to bottom. (Drawings shown later in this chapter).

Remember that the Energy String is really a group of Energy Wavelettes, like Energy Fish. Therefore two groups of these Energy Fish can merge, as long as the individual fish are traveling in the same direction. The result is a larger group of energy fish, and therefore a larger Energy String. This increase in size can either be in thickness or in length.

Background Concepts

Overview

There are a few background concepts to keep in mind as we discuss the flow directions of magnetic energy strings. Some of these have been discussed before, but are worth highlighting again.

1. Magnetic Energy Units, as Magnetic Energy Fish

Remember that a magnetic energy "string" is really a collection of smaller magnetic energy "units". We often call these Magnetic Energy Fish. This is because these energy units will group together and swim together in the same way that fish of the ocean will do.

Therefore: the direction of the string as a whole is the aggregate directions of the energy units. For example, when each of the energy fish units travel northeast, then the string as a whole will travel northeast.

The important thing to notice is that each of the individual energy fish units can flow independently; and yet at the same time, these energy fish units are flowing together in a cohesive group.

Most of the time these energy units travel in the same direction. This becomes the primary flow direction.

However, the individual energy units within the group can change direction. This can cause the string (group of energy fish units) to change the overall direction, and sometimes to completely reverse the flow.

2. Gravity Strings Influence Direction

The Gravity Strings which are attached to the Magnetic Energy String have an influence on the directions. We will see how the direction of pull of the Gravity Strings can induce the flow in the Primary Direction. Similarly, a different direction of pull of the Gravity Strings can induce the nearby energy units to flip direction; ultimately causing Reverse Flow of Magnetic Energy.

3. Merging only if Aligned

The merging of any two energy strings will occur only if the two energy strings are aligned, flowing in the same directions. The strings can either be side to side alignment or top to bottom alignment. In both cases, the individual energy wavelettes of each string are moving in the same direction.

Primary Flow Direction

A group of fish in the ocean will generally travel in the same direction. Although each fish is an independent entity, they choose to swim in the same direction when traveling as a group. Therefore, the Primary Flow Direction of this group of fish is the same direction as all the individual fish are swimming.

The same is true for our Magnetic Energy String. The "string" is actually compose of smaller energy units. Each energy unit (energy fish) can travel independently. Yet most of the time, when grouped together, these energy fish will travel in the same direction. Therefore the overall direction of the Energy String is the same direction as all of the individual energy fish are traveling.

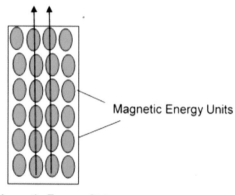

Merging of Magnetic Energy Strings
If Same Flow Directions

Overview

We have said many times that Magnetic Energy Strings will merge together…if the strings are flowing in the same direction. This process is similar to having two groups of fish, flowing the same general direction, merging into one larger group.

Now we are ready to consider the ways in which two energy strings may merge together. The two most common will be side to side merging (which makes a thicker energy string) and top to bottom merging (which makes a longer energy string.

In the following sections we will illustrate more precisely the physical reality of the Magnetic Energy Strings merging. As we proceed with the illustrations, remember the following:

1. Gravity Strings, which are attached to each Magnetic Energy String, will intertwine and pull the respective Magnetic Energy Strings close enough to touch. This allows the Magnetic Energy Units to merge.

2. The Magnetic Energy Fish Units must be flowing in the same direction for actual merging to occur.

3. Magnetic Energy Strings can merge either in thickness or length.

A. Side to Side Merging of Magnetic Energy Strings

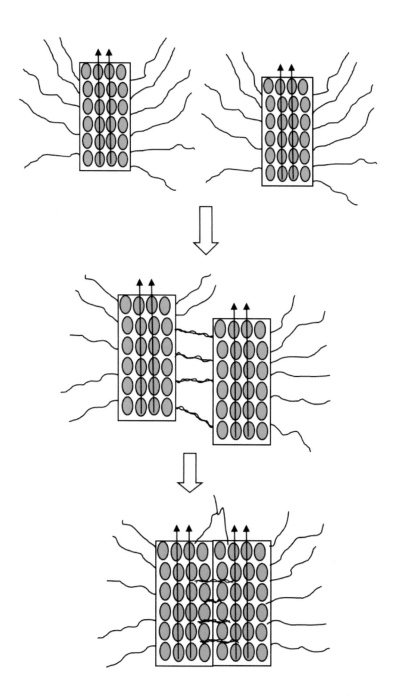

Notice that the Magnetic Strings which are side by side, are pulled together by the mutual pull of the respective gravity strings. Notice also that when Magnetic Strings are physically merged, some of the mutual pull gravity strings are embedded, while others merge in the Y shape.

B. Top to Bottom Merging of Magnetic Energy Strings

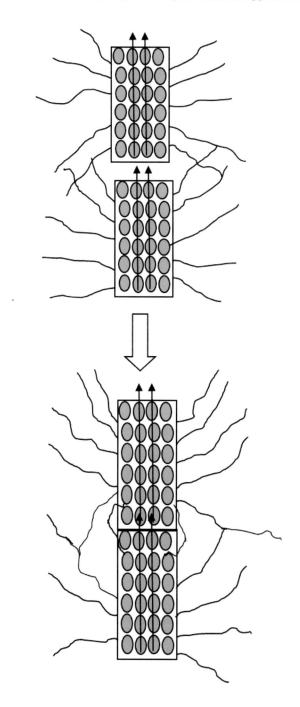

    Notice how the bottom Magnetic Energy String easily merges with the top one. This is how Magnetic Strings join to become longer. We also see this in the traditional magnets snapping together.

    Also notice the gravity strings which brought the two together have themselves merged, and created the extended Y shape.

Changing Flow Directions: As Group, and as Individuals

A more advanced topic which we will discuss in this chapter is the changing of direction of the Energy. This is analyzed in two ways: 1) the changing direction of the energy *string* (group as a whole), or 2 the changing the direction of the energy *wavelettes* (individual units). The processes are different.

Briefly: the group as a whole (string) will swerve left and right. This will occur on its own, or when encountering an opposing group. However, the string (group as a whole) will not completely turn around.

However, the individual units within the string can change direction. These individual energy wavelettes can act independently of the others. Any one or more of these wavelettes can pivot its traveling direction, reverse its swimming direction, or leave the group completely. These actions are usually induced by the influence of nearby gravity strings.

Note that under normal circumstances the majority of energy wavelettes within a group are flowing in the same group. Therefore the group as a whole will travel in that direction.

Gravity Strings Induce the Direction of Flow in Magnetic Energy Strings

Introduction

In earlier books we stated that Energy Strings flow only in one direction. However, this is not entirely true. Under certain circumstances, the energy strings can change their direction of flow.

In general, the direction of energy flow for an energy string is determined by the pull directions of the gravity stings. In this section we will discuss these concepts in detail, and we will see the numerous practical effects.

Gravity strings induce the direction of energy flow for the energy strings.

One of the most important concepts regarding direction of flow for energy strings is this: Gravity strings induce the direction of energy flow for the energy strings.

For example, if the gravity strings pull upward on the magnetic energy string, then the energy flow of the magnetic energy string will also be upward. If the gravity strings pull downward on the magnetic energy string, then the energy flow of the magnetic energy string will be downward.

Primary Flow Direction: Momentum of the Magnetic Energy Units

The pull of the gravity strings will start the regions of magnetic fish moving in a particular direction. Yet once these magnetic fish start moving…there is almost no stopping them!

The energy within our energy string is much, much, much greater than any set of nearby gravity strings. Therefore, once the energy string starts flowing in one direction, this flow stays that way for a long time, regardless of what the gravity strings may do.

Once we start the energy string flowing in one direction, that is the direction which the energy string flows. This direction becomes the Primary Flow Direction. There is so much energy in that energy string, that there is a type of momentum. Thus, the flow tends to continue in that one direction. This is why we tend to see energy strings flowing in one direction, for much of the time.

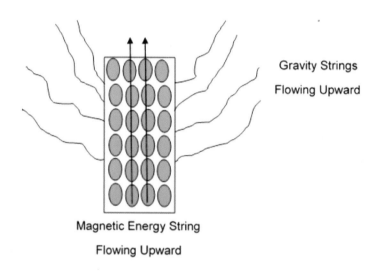

Gravity Strings Flowing Upward

Magnetic Energy String Flowing Upward

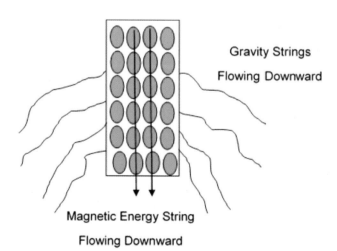

Gravity Strings Flowing Downward

Magnetic Energy String Flowing Downward

Momentum is Difficult to Stop

Note also that the gravity strings may wave about, and will often intertwine with other objects, yet not influence the flow direction. It takes much more than that to change the direction of the magnetic energy units. (This is why we can safely say that most of the time, the flow will continue in one primary direction; with only minor shifts from passerby energies).

Thus, the gravity strings may find another object and intertwine with it. Yet the energy in our energy strings are still flowing in the same direction, regardless.

Think of it as a horse: the horse is running, at fast gallop, and it will take a lot of pull on the reins before the horse slows down and changes direction.

A More Advanced Model of Energy String and Gravity Strings

This leads us to a more advanced model of energy strings. In the diagram below we have the same basic situation: the majority of gravity strings flow upward, which induces the energy fish in the energy string to flow upward.

We now add gravity strings which point downward. These downward pointing gravity strings may induce the nearby magnetic fish to also pivot, and point downward.

However, the overall flow of the energy patches in the magnetic energy does NOT change, due to the momentum described above.

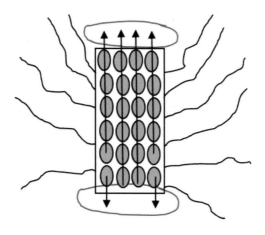

Different Flow Directions Within an Energy String

If each gravity string affects only the nearby energy units, does that mean that different regions can flow in different directions? The answer is yes. Therefore, indeed, different regions of an energy string can be flowing in different directions at the same time. However, the energy units will tend to stay together. Note that this concept will also help us understand how the change in direction of energy flow can occur.

Induced to Change Direction, Though Slow to Change Direction

The direction of energy flow is steady, until induced to change. Once the direction of energy flow has been set, this is the direction it will be, forever, until the gravity strings change direction. Thus, *free flowing* energy strings in space will continue to flow in their directions forever.

However, magnetic energy strings which are gravitationally connected to other magnetic energy string may in fact cause its partner to change flow direction.

This is because the gravity strings which are intertwined provide a pull in a particular direction. This may induce the region surrounding the mass spot to change flow direction. Therefore, as the two magnetic string are pulled together, it will sometimes be the case that some of the magnetic fish are induced to pivot where the gravity string pulls, and change flow direction.

For example, if the original magnetic string flows north, but some of the gravity strings pull southwest, then the region surrounding those gravity strings may also pivot southwest. Then, as more gravity strings are pulling southwest (because they are intertwined with another object over there) then many of the magnetic energy units may be induced to flow southwest.

In a gradual process, the magnetic energy units start pivoting from flowing north…to southwest. The pivoting is further induced by the momentum of the already pivoted energy fish. Eventually the majority of magnetic energy fish have pivoted in direction.

This is the basic cause of how magnetic energy units can be induced to change flow direction.

Pivoting Occurs for Both Magnetic Energy Strings

Note that this pivoting operation will usually occur for BOTH of the Magnetic Strings. Thus, when both of these magnetic energy strings do physically touch, they WILL be flowing in the same direction. This will allow them to merge.

This is the process which occurs in the EM Pulsation process. When the magnetic strings contract together, they are not only pulled together, both will also reverse their flow. Therefore, when the two magnetic strings merge, they will now be flowing in the opposite direction.

Energy Flow Direction of EM Strings in Pulsation

Introduction

What we know of as "Frequency" is actually a pulsation process. The energy strings which ride the photon core engage in a type of pulsation. The rate at which this pulsation occurs is the frequency we measure.

The pulsation process involves the energy strings expanding outward, contracting inward, and then contracting outward again – in the opposite direction. This process repeats forever.

The pulsation process can be divided into different steps, and there are different causes for the three different processes:

1. The outward explosion is caused by mass-energy conversions.

2. The inward contraction is caused by mutual gravitational pull of all the gravitational energies. (These gravity strings are on the particle as well as on each of the magnetic strings, ultimately causing a contraction).

3. The change in the direction of pulsation is ultimately related to the internal flow of the energy units within each string.

All of these steps will be discussed in great detail in the book *Photons in Motion*. Shortened versions will be found in the next two chapters.

Chapter 4
Mass-Energy Conversions

Chapter taken from book *Photons in Motion*.

Introduction

Topics Explained

In this chapter we will explain the conversions between mass and energy. More specifically, this means we will discuss the following:

1. Conversion of Energy to Mass
2. Conversion of Mass to Energy
3. Re-Conversion of Energy to Mass, using Protomass

These concepts will be explained in physical terms (not mathematical) and will be illustrated in detail.

Physical Realities of Mass, Energy, and Conversions

It has long been known that mass and energy are variations of the same thing, however there has been a lack of understanding of the physical realities. We will here provide the physical explanations of energy and mass.

Similarly, there are equations which help convert the values mass to energy. However, scientists have lacked a visual understanding of the physical processes. In these chapters we will explain and illustrate these conversion processes.

Note that we will mostly focus on the conversion of mass to energy. We should emphasize that the internal explosion, due to mass-energy conversion, is a very significant component in the EM pulsations process. This process is also essential for the fast speed of the photon ("speed of light").

We will also spend time discussing the reconversion of energy into mass. This re-conversion is also important for the pulsations process.

Importance and Applications

There are several applications of mass to energy conversions for the motions of photons. This includes: 1) speed of the photon core (the famous "light speed); and 2) the pulsation frequency. These applications will be discussed in great detail in the next chapter.

Mass to Energy Conversions: Basic Concepts

Overview

Scientists have long known that mass and energy are both forms of the same thing. However, they lacked a physical understanding of what energy and mass are, as physical entities.

Furthermore, scientists have created various equations to estimate the amount of energy in mass, yet lack the understanding of how the physical conversions will occur.

In this section we will provide a brief physical description of the entities and conversion processes. These will be explained in greater detail in subsequent sections.

Energy Units

Energy exists as small units of energy. We can call these entities Wavelettes of Energy. They are tiny entities, which move similarly to birds or fish. These energy units can also be called Energy Fish, as they swim and move together in groups as the fish in the ocean.

Types of Energy

All Energy Units are ultimately the same thing: the Universal Energy. The difference between each type of energy is the amount of Universal Energy in the wavelette, and the geometrical structure of the wavelette.

This is similar to origami, where paper is folded into different complex shapes. Using larger paper, and forming into a more complex structure, will create a larger type of energy unit. Thus, using more of the Universal Energy, and forming into a more complex shape, creates a stronger energy wavelette. This is the case for the Magnetic Energy Wavelette.

Conversely, less amount of the Universal Energy, and similar structure, creates a weaker energy wavelette. This is the case for the Gravitational Energy Wavelette.

*These concepts will be expanded in greater detail in the upcoming book on the Theory of Everything.

Mass Spots

Mass is simply concentrated energy. This mass exists in Mass Spots, which are similar to an oil spot. Pressure from outer wavelettes push on the inner wavelettes, to create a concentrated spherical entity. This is the mass spot.

Mass spots can vary in size, depending on the amount of energy compacted. However, even the smallest mass spots will have an enormous amount of energy within.

Furthermore, as there are different wavelette types (the types of energy units) the mass spots will necessarily be composed of different types of energy. Yet, ultimately these are all the same, as everything is composed of the Universal Energy. The difference (whether wavelette type or mass spot type) is in the amount of Universal Energy, and the specific geometric structure created.

Energy to Mass

The conversion of energy to mass is done by external pressure, similar to pressure on gas creating liquid. The outer energy wavelettes push on the inner energy wavelettes. This pressure condenses the inner energy wavelettes into a more viscous substance. The energy wavelettes are not only compacted, but interact and create a new (and highly complex) physical structure. This structure is the Mass Spot.

Mass to Energy

Mass spots can be converted back into energy wavelettes given the right circumstances. This is usually done with extreme pressure.

Notice that mass spots are created by pressure, yet additional pressure will also explode the mass spot into pure energy.

In the beginning of the mass formation, the outer energy wavelettes apply enough pressure onto the inner wavelettes, to condense the inner wavelettes into the mass spot. At this point, the mass spots will generally remain in place forever. The existing pressure is not enough to burst the mass spot. There is a status quo, where the formed mass spots will remain in place.

However, under the right circumstances, there can be additional pressure from the outer wavelettes, forced onto the inner mass spot. The mass spot does not condense any more, therefore the pressure will only serve to explode the mass spot.

The pressure from the outer wavelettes is so strong that the mass spots are not just condensed, but exploded outward. This is similar to the explosive release of a volcano. At this point, the mass spot is reconverted back into free-flowing energy wavelettes.

Occurrences of Mass-Energy Conversions

Most of the time, mass spots remain as mass spots, and energy portions remain as energy portions. However, there are some times when partial or full conversions will take place.

Note that these are important for several Photon Motions

1. Interior of a photon: Full conversion from mass to energy.

2. Pulsation Strings: Partial mass to energy conversion, followed by rebuilding of mass spot.

3. Rearranging energy strings: low level mass to energy conversions will cause energy strings to break into segments.

4. Launching energy strings: sometimes a few pockets of mass to energy conversions will launch individual energy strings from a particle (free strings, not necessarily as an EM burst). Similarly, this process may also induce launching of a photon system.

Partial or Complete Mass-to-Energy Conversions

Mass to energy conversions can be partial or complete. If partial, then there will still be some mass spot left; the mass spot will simply be much smaller in size. The process may also be close to complete conversion but not quite complete. This will leave a "protomass" in place (see below).

A complete mass to energy conversion will produce trillions and trillions of energy portions. The number of energy portions created from one basic unit of mass spot is enormous.

Re-Conversion of Energy into Mass

Mass spots can be created and rebuilt at any time, given the right circumstances. However, some mass spots are more quick to re-build than others. They key is the amount of mass left behind in the partial conversion.

If there is plenty of mass remaining in the mass spot (just reduced in size), then this mass spot can be rebuilt to normal size quite easily. This happens quite often, particularly in the explosions of the EM pulsation. The explosions are enough to send the field strings outward, yet enough mass left intact for the full mass to be reconstructed quite quickly.

Using Protomass to Rebuild Mass Spots

The minimum requirement for rebuilding the mass spot is the "Protomass". As long as this protomass exists, and has not been destroyed, then the mass spot can be rebuilt easily.

This protomass is not true mass, but it is glob of substance which is essentially "energy-mass". In the Protomass substance, the portions flux between the structure of mass and the structure of energy.

The Protomass can be considered like the foundation of a house or as seed for crystalline growth. With this Protomass in place the full mass spot can be rebuilt in a short time. This is important for the fast rate in which the field strings can undergo repeated explosions in field pulsation.

Next Discussions

Now that we discussed the main concepts related to energy, mass, and the conversions between the two, we can proceed with more expanded discussions on each topic. We will focus most on the following:

1. Conversions of Mass to Energy, and resulting Explosions

2. Types of Explosions

3. Rebuilding Mass Spots using Protomass.

Mass Spots Converting to Energy Portions:
Overview of Process

Mass-to-Energy Conversion from Additional Pressure

Mass spots can convert into energy. More specifically, this means that any one of the Mass Spots will convert into additional Energy Portions.

The basic mechanism is what we will refer to as the "pressure cooker" effect. The outer energy portions apply pressure on the mass portions. This pressure begins another mechanism which breaks apart the mass spot structure, and therefore releases the energy wavelettes.

The key to the process is "pressure" from the energy portions which surround the mass spot. When there are enough energy portions applying pressure to a particular mass spot, this will induce a secondary process (yet to be determined) which converts the mass spot into energy.

Pressure to Build, Pressure to Break Apart

Notice that pressure is required to build, yet additional pressure is required to break apart.

We see in this in many areas of science. For example, when we press two protons together they can fuse together as a nucleus. Yet we send the protons together at high speeds, and they will break each other apart.

A similar concept exists with the building then breaking apart of the mass spots. We require enough external pressure of the outer wavelettes on the inner wavelettes to compress the inner wavelettes into mass. The energy wavelettes reconfigure into a new complex structure, which is the mass spot. Yet we apply enough external pressure, then the mass spot structure will break apart, and the wavelettes will be released.

Stability of Mass Spots

Note that most of the Mass Spots are stable. In most magnetic and electric strings, whether field strings, driver strings, or the particle structure, will have mass spots which remain in place. This is because there is usually not enough external pressure to break the mass spot.

Furthermore, when the Magnetic Strings break apart or merge together, the mass spots will generally remain as they are. The mass spots continue to exist as they are. They seem almost permanent.

However, when magnetic strings come together at higher velocities, or at the precise angles, then there will be enough pressure to break apart the mass spots.

Interior Mass Spots, Thicker Strings, and Missile Shots

It is also for this reason that most of the mass-to-energy conversions will come from mass spots which exist in the interior of the energy string.

Furthermore, thicker energy strings are more likely to undergo mass to energy conversion. This is because thicker energy strings have more energy portions than thicker strings, and thus are capable of applying a greater total pressure on any internal mass spot.

In addition, a strategic missile shot of energy can break apart the mass spot. When a secondary magnetic energy string comes in to the first magnetic string, from the side, and hits the mass spot directly that incoming energy may break apart the mass spot.

Mass-to-Energy Conversion Options

Because this "pressure" mechanism is required for conversion, the mass portions will be converted to energy portions only if:

a) There is significant pressure from the surrounding energy portions on that mass spot to induce conversion.
*This is the case for thick strings with interior mass spots.

OR

b) Two energy strings collide at a very fast rate.
*This is what happens when pulsation strings collide and cause an explosion. The fast moving energy string applies great pressure from various energy portions all at once. This pressure will be applied on the mass spots which contain the gravity strings, of both energy strings.

Mass Spots Convert to Enormous Amounts of Energy

$E=mc^2$

The amount of energy converted from one unit of mass is enormous. According to Einstein's famous equation, energy is related to mass as $E=mc^2$. The multiplier here is "c^2", which is a value of 9×10^{16}. This an enormous number.

When we apply this to our new understanding of energy strings, we can see that one mass spot will be converted into a vast amount of energy. This means that the smallest possible mass spot, when converted into energy portions, will be 9×10^{16} individual energy units!

Imagine having a single stone in your net, the snapping your fingers, and having 9×10^{16} individual fish appear in your net. That is essentially what we have going on.

Note that this enormous amount of energy will have significant effect on the overall motions of the photon system, including the speed of light.

Different Size of Mass Spots and the Amount of Energies

Furthermore, mass spots will exist in different sizes. This means various amounts of energy wavelettes were compacted and rearranged into that mass spot, and therefore can be released when the mass spot is broken apart.

As an advanced concept, note that there are several equations which relate mass to energy. The equation of $E=mc^2$ is only the most famous, and this isn't even the full version. The variations in equations is due to the various circumstances in physical science. These physical realties, ultimately, are related to the size of the mass spot.

Thus, depending on the location of the field or the type of the particle, there will be different mass spot sizes and different mass spot arrangements. Therefore: the exact conversion value between mass to energy can vary, depending on how much of the original wavelettes were compacted into the mass spot.

In any case, the multiplier of "c^2" is a good approximation; letting you know that in most mass spots there is a significant quantity of energy units compacted within.

Also remember that mass spots can be partially converted, with only some of the mass broken apart. Yet the amount of energy released will be significant in any case.

Example of Mass–to-Energy Conversion Resulting in Explosion of Energy

As stated earlier, one mass spot will create an enormous amount of energy. The amount of energy produced from one mass spot is amazingly, ridiculously enormous.

Thus, when one mass spot transitions into energy portions, we will see that trillions and trillions of energy portions are produced. This is the essence of the "internal explosion" from within the energy string.

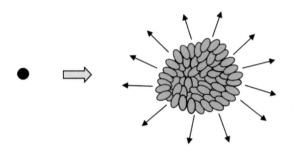

Mass-Energy Conversion From Within a Thick String

Introduction

Mass spots can be converted to energy units under the right circumstances. The mass-to-energy conversions will most commonly occur in the following circumstances:

1. A single thick energy string has enough internal pressure to cause a conversion from within.

2. Two energy strings collide, in such a way as to induce the mass spot to break apart into energy.

In the following sections we will look at the first situation: a magnetic energy with such thickness that the mass spots will have been created, and then further broken apart.

Mass-to-Energy Explosion

Both cases above will create a type of explosion. We will first see this in where a thick string induces its own mass spots to explode.

When enough pressure from the Energy Wavelettes is applied to a Mass Spot, this Mass Spot will convert into additional Energy Wavelettes. Notice that there will in fact be trillions and trillions of energy portions released from one, single, mass spot. This is what causes the mass-energy explosion (as illustrated above).

This vast creation of energy, this internal explosion, will then be powerful enough to blast apart the one energy string into several smaller segments. The process is illustrated below.

Drawings of the Mass-to-Energy Explosion from Within a String

The diagrams below illustrate a thick magnetic string, which will break apart its mass spot from within. This thick magnetic string has enough internal energy to create the mass spot, then later to induce the mass spot to break apart.

For simplicity in the drawings below the gravity strings are not shown; nor are the mass spots on the edge, as these are not involved in this particular process. However, in the pulsation process (discussed at the end of the chapter) the outer mass spots and the gravity strings *are* absolutely part of the process.

In the set of diagrams below you will see the following steps occur. Note that each step is numbered in the illustration.

1. Normal Status for the Energy String

2. Mass to Energy Conversion and Resulting Explosion

3. Energy Wavelettes Push Outward

4. Energy String Breaks Apart into Smaller Segments

1. <u>Normal Status for the Energy String</u>

In the first diagram we see the magnetic energy string as it normally exists. The majority of the energy string is composed of energy wavelettes. These are represented by the green ovals. Notice that the shape of the oval indicates the direction of the energy flow. In diagram #1 below, all energy wavelettes are flowing to the right.

In addition to the energy wavelettes there are a few mass spots. These are represented by black circles. Note that the mass spot which undergoes mass to energy conversion can be anywhere, however the most efficient and complete mass to energy conversion will occur for mass spots in the center of the energy string.

Note also that only one mass spot is shown. This is for simplicity in discussing this process. Many more mass spots will exist in reality.

2. <u>Mass to Energy Conversion and Resulting Interior Explosion</u>

The mass spot is converted into energy wavelettes. The exact process is not known at this time. However, I believe it to be a type of pressure cooker mechanism where the surrounding energy portions apply pressure to the mass portion, causing the mass spot structure to break apart. This will convert the mass spot back into individual energy wavelettes.

What *is* known is that one unit of mass will be converted into trillions of energy wavelettes. The resulting effect is an internal explosion, where these trillions of energy wavelettes are released all at once.

3. <u>Expansion as Energy Units Push Outward</u>

The trillions of energy portions which are released from mass to energy conversion will naturally expand outward. These released energy units will then push on the pre-existing energy portions. This results in those surrounding energy units being pushed outward in various directions. This is like a balloon being expanded from the inside. The net result is that all energy units – the preexisting and those from the mass spot – will create an expanded energy string.

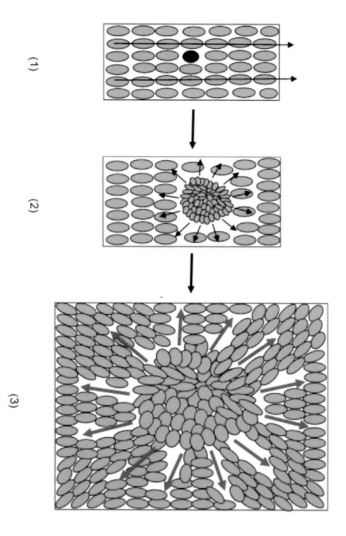

4. Energy String Breaks Apart into Smaller Segments

The last step in the process is where the energy string breaks up into smaller segments. This occurs because of the immense internal energy (from internal explosion) pushing the energy wavelettes in all directions.

Earlier we said that there are also micro gravity strings which connect the individual energy wavelettes together. This is how most of the energy portions remain "together" as one cohesive energy string.

Yet, if we provide this internal explosion (due to mass-energy conversion) this will push the pre-existing energy wavelettes. The "push" is the key concept here. These energy portions are pushed outward, and at a fast rate. This can cause many of the small gravity strings connecting the energy portions to separate from others.

Thus, due the force of the force of the internal explosion; and the separation of micro gravity strings, groups of energy wavelettes will begin to separate from the main string. In other words: a smaller energy segment has broken off from the main group.

Furthermore, because the mass-energy explosion pushes out energy wavelettes on all directions, the there is a strong push of the wavelettes in all directions. This can result in several Energy Segments breaking off from the original Energy String.

Directions of the Energy Segments

Notice also the directions of these new segments. The explosion of energy will likely have many more wavelettes than the nearby region. This means the directions of each segment which is broken off will be determined by the direction of the exploded energy units.

Where the exploded wavelettes push in the same direction as the original energy string, then the wavelettes will continue that direction, but much faster.

However, where the exploded wavelettes push away from the original direction, these wavelettes will force the existing wavelettes to change direction. The segment will therefore be blasted away in that direction.

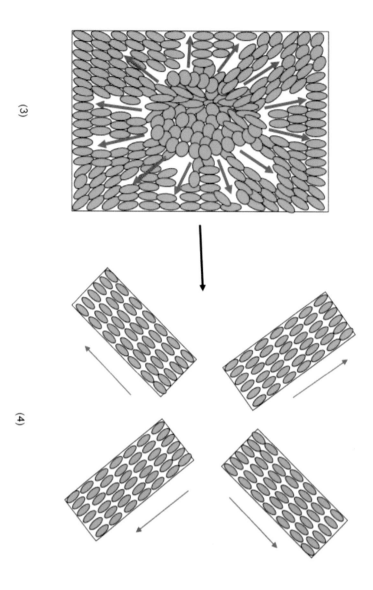

Partial or Full Energy String Breaking

Note that this process can be partial or full explosion. If partial, then only one or two small segments will break off. If full, then the entire energy string will break apart into numerous smaller segments. Then of course there are ranges of segment breaking options, depending on the amount of energy released from the mass spot.

Whether the mass spot conversion will be partial or full, and the resulting size of the individual segments, will depend on three factors:

a) the size of the mass spot before conversion,

b) the size of the original energy string, and

c) the amount of pressure which energy wavelettes apply to the mass spot.

Advanced Note on Wavelette Directions Repelling or Not

*Note that the act of "repelling" of wavelettes is dependent on the relative force of the wavelettes moving in each direction. This essentially means the amount of wavelettes facing each direction, in that region, at that moment.

a) If the incoming force of energy wavelettes is very weak there will be no repelling; the opposite flowing strings can be aligned. This is the situation for the Gravity Strings, which are weak enough to slide alongside each other.

b) Where the incoming force is approximately equal (which is the case most of the time), then there will be repelling. This is where we see each string pushing each other away, or bending their tips as they pass.

c) However, when one group of energy units has a much greater force than the others, this can cause the weaker group to turn and adopt the stronger direction. This is the case in the mass-energy conversion explosions.

Traditional Explosions versus Conversion Explosions

Two Types of Explosions

Before we continue, we must clarify a very important point regarding explosions. There are two main types of explosions. The first type is a release of energy strings. The second type is the release of energy wavelettes from the mass-energy conversion. This clarification is important for understanding of many areas of science.

Notice that the majority of explosions we observe involve the first type, where there are energy strings are released. Yet in these cases there is no mass-to-energy conversion involved; the mass remains intact.

However, most of the topics in this book involve the second type of explosion. This is where the mass spots are converted into wavelettes of energy, which then results in an explosion among the energy strings.

1. Most Explosions are Release of Energy Strings

The majority of explosions involve the release of energy strings. Similarly, most cases of energy transfer involves the relocation of energy strings. In these cases the energy strings are released; yet the mass remains intact. It is the great number of energy strings released, at the same time, we have created an "explosion".

Remember that all things are composed of Energy Strings. Each particle and each energy field is composed of magnetic or electrical energy strings. The internal energy which propel objects forward are energy strings. Furthermore, all molecular bonds are energy strings.

Indeed, every object and every motion can ultimately be traced back to the magnetic and electric energy strings. Therefore the transfer of energy is always a transfer of energy strings.

Thus: most explosions are in fact the release of these energy strings. The release of energy, whether from inside a particle or the breaking of a molecular bond, is always the release of energy strings. Released in small amounts, we explain as "heat". Yet, whenever many of these energy strings are released from their location, at the same time, this is what we know of as an "explosion".

Regardless of how the particular explosion is explained in the texts, most of these explosions can be seen as the sudden release of numerous energy strings. For most explosions, the process is where numerous energy strings are thrown into the air at the same moment.

2. Mass-to-Energy Explosions

The second type of explosion is the conversion of mass into energy. An individual mass spot can be broken apart, and the individual energy wavelettes released. Then, because there were so many energy wavelettes packed into that mass spot, the sudden release of wavelettes creates a type of explosion among the energy strings.

This process is used in many areas of photon motion, and therefore will be the explosion type discussed in this book.

Differences Between the Two Types of Explosions

The primary difference between the two types of explosions is the entities involved: energy *strings* vs energy *wavelettes*.

In the traditional explosion, we are looking at the energy *strings*. In this type of explosion, we have numerous energy strings being released at the same type. Of course each energy string is actually composed of wavelettes, but it is the ejection of the many *strings* which creates the energy burst.

In contrast, for the mass-to-energy explosion we are looking at the energy *wavelettes*. The burst of energy comes from the mass spot, which means the sudden release of *wavelettes*, which is different from the sudden release of strings.

In fact, the conversion of mass to energy will involve very few strings. Only one string is required, though usually the collision of two strings is most common, to induce the conversion process. Thus we have only two strings in the mass to energy conversion explosion, rather than the many strings involved in the traditional explosion

Furthermore, the amount of energy created from the conversion is enormous. Therefore, one unit of mass will create an insane amount of energy in the conversion process. This is enough to cause the string to explode apart into smaller segments.

Importance of Type 1, but Focusing of Type 2

In traditional explosions the strings themselves remain intact and are thrown into the air as they are. It is important that you are aware of the basic process, since the majority of other explosions are based on that process. However, we will not be discussing traditional explosions much in this series.

In this book, and through most of the series, we will be discussing the details of mass to energy conversions, and the resulting explosion from within the energy string. In these types of explosions the individual strings themselves explode, due to release of energy wavelettes.

Thus, the Type 2 Explosion is a "Conversion" Explosion. In these explosions some mass of the energy string is converted into energy wavelettes, within that same string. This conversion will eject additional energy in great quantities. This action is therefore an explosion within the energy strings themselves.

Also note that this process is important for the speed of light and the pulsation frequency.

Protomass: The Transition from Energy to Mass

Protomass Overview

Mass is concentrated energy. The energy wavelettes are pressed together with enough force to create a new geometrical structure. This new structure is the Mass Spot.

However, there is a transitionary phase between energy and mass which we will call the "protomass". This protomass is not true mass, but it is a flux between the structure of mass and the structure of energy. This protomass is semi-liquid, semi-solid substance, something like glue.

This protomass acts as a seed crystal, where the presence of this protomass makes it much easier for other energy portions to join the system. This protomass is therefore an essential transition entity to building a true mass spot.

In the following sections we will explain the Protomass in detail, as well as how the Protomass will assist in creating the true Mass Spot.

Conversion of Energy into Mass

We know that energy can be converted into mass. We will discuss the process in this section.

Although the process is not completely understood, a key component is what we call the "Prototype Mass" or "Protomass". This "protomass" is an early formation of what will become the "mass spot" discussed earlier. Without this protomass, it is almost impossible to create a true mass spot.

1. The process begins with the inner wavelettes being pressed together strongly enough to create the protomass. In this form, the wavelettes fluctuate between the structure of the mass spot and the structure of the wavelettes.

2. The Protomass acts as a type of seed crystal. With the presence of the Protomass, more inner wavelettes, and pressure from the outside, more of the energy wavelettes will join the structure of the protomass. This will gradually build up the Protomass into a true Mass Spot.

Physical Properties of the Protomass

The physical properties of the protomass are a hybrid of the properties of pure mass and pure energy. In the "protomass" the energy portions are converting into mass, yet they are also converting back into energy again. The mass-energy portions are in a state of flux, sometimes as mass, sometimes as energy. They are never a true mass (which is more permanent), yet they never leave as pure energy.

Notice that whenever one thing changes into another, there is a transition stage. The "protomass" is an example of a group of entities which are in that transition stage.

We can use the analogy of steam, water, and ice. Steam and ice are two different physical forms of essentially the same things. Yet to convert steam into ice we must go through transition stage. This transition stage we know of as "water".

Thus, I believe the "protomass" to be a group of energy-mass entities. These entities are not only grouped together, but are in the structural transition stage, between the form of energy and the form of mass.

Protomass as Seed Crystals

We can also compare the Protomass to the use of Seed Crystals. In chemistry and cooking, when we want to form certain crystals it is best to begin with seed crystals. These seed crystals assist the building of crystals tremendously. Protomass acts in the same way.

It is well known that it is difficult to get atoms to come together in a crystal form. Most often the atoms and molecules will move about, but not come together. Molecules in solution (floating in some liquid) will move around each other, bounce off of each other, yet not always form a bond.

This process can be assisted with the seed crystal. There are many applications in chemistry and cooking where we begin with a crystal, already formed and stored just for this purpose. This is our seed crystal. We put the seed crystal in the solution, and now the other molecules have something to latch onto. The other molecules in solution will now attach to the crystal, one at a time, gradually building the crystal to a larger size.

Thus, the seed crystal is often used to encourage molecules in solution to join that crystal, which then creates larger and larger crystals.

This "seed crystal" process is essentially what our "protomass" does. Our protomass acts as a seed crystal. With the existence of the protomass, nearby energy portions are encouraged to join the protomass. (This is similar to how the seed crystal encourages atoms in solution to join the existing crystal).

And just as the crystal grows larger and larger as molecules in solution join the crystal, in our energy strings the protomass grows larger and larger as energy portions join the protomass system.

Mass Spot Creation Rate Depends on the Existence of Protomass

The rate of creation of the mas spot depends on the existence of the protomass. If there is a protomass, then a true mass spot can be created. However, if there is no protomass, then the protomass must be created first, before the building of the true mass spot.

*This is important for the rebuilding of the mass spot in the EM Pulsation Process. There is always some mass left after the explosion, which is at the very least the presence of the protomass. Therefore, the energy wavelettes can be compacted again rather quickly, because of the presence of the continued presence of a protomass.

We can again compare this to our seed crystals. Without a seed crystal the molecules in solution tend to stay in solution. They rarely form any crystals. The same is true for our mass spots. Without any protomass nearby, the energy portions will keep moving about, and yet never come together and transition into a mass spot.

Conversely, if we have a protomass then the nearby energy portions are more inclined to join that protomass. The existence of the protomass encourages other energy portions to join. There is a place, a mechanism, for the energy portions to join together.

Furthermore, just as the crystal grows and grows due to additional molecules joining the existing crystal, our protomass can grow in size due to additional energy portions joining the existing protomass. The existence of the protomass, like the seed crystal, encourages energy portions to join. This allows the protomass to grow in size.

Conversion from Protomass to True Mass Spot

At some point, when there are enough energy wavelettes attached to the protomass, then the final conversion process begins. That is, for the full conversion of energy portions into a single mass spot it is required that a certain number of energy portions are attached to the protomass. Then and only then will the final and full conversion take place, thereby creating a true mass spot.

The exact conversion process from mass to energy is still to be discovered. However, in basic terms the mass spot it "converted" from protomass to true mass. This means that:

a) A minimum number of energy wavelettes have been pressed together.

b) The energy wavelettes are now interconnected together.

c) The geometric structure of the grouping of wavelettes has reconfigured to become a true mass spot.

What we do know is that it requires an enormous amount of energy to become one unit of mass. This means that trillions of energy portions must somehow be "reassembled" into what we know of as mass.

Importance of Critical Protomass Size to become Mass Spot

A true mass spot will be formed when the protomass reaches a critical size. Only when the protomass reaches this size will the actual conversion process from Energy to Mass proceed.

When the energy wavelettes are still in the "protomass" phase, they are not truly mass. Some energy portions will certainly be close to mass form, yet other energy portions will be closer to energy form. This is the situation we will have until the protomass reaches a type of critical size. Only when enough energy portions have joined the protomass with the conversion process take place.

This is an important concept. The process requires a certain amount of "energy" for the conversion process to take place, or certain number of energy wavelettes interacting and bonding in a certain way, before the whole protomass can become the full and complete mass spot.

Geometrical Structure

The exact geometrical structure of the mass spot is yet to be determined. At the moment, I believe it to be a highly complex geometric shape. I believe the wavelettes to be joined, merged, and weaving among themselves. In essence, the mass spot is tiny version of the neutrino; a near solid sphere of complex geometrical pattern.

Review of Mass-Energy Conversions

Let us now review the main concepts of Energy, Mass, and the Conversions between the two.

1. Energy exists as wavelettes. These wavelettes move about independently, yet often travel in groups.

2. When energy wavelettes travel in a group, we refer to this as an "energy string".

3. There are three main types of energy: Magnetic, Electric, and Gravitational. Each type of energy exists as its own type of energy wavelette; and the same types of energy wavelettes will create each type of energy string.

4. Ultimately, all type of energy are variations of the Universal Energy. The difference between each type of energy is the amount of the universal energy in the wavelette, as well as the geometric structure of the wavelette.

5. Mass is concentrated energy. Mass exists as Mass Spots, which is a complex geometric structure of highly packed energy wavelettes. This mass spot is created by pressure of the outer wavelettes on the inner wavelettes.

6. The transitionary stage between true energy and true mass is the Protomass. The Protomass is a hybrid of mass and energy. The entities are in state of flux, sometimes in the form of mass, sometimes in the form of energy.

7. The Protomass acts as a seed crystal. With the existence of the protomass, it is easier for other energy wavelettes to join the geometric structure of the mass which is being formed.

8. When the protomass has enough energy wavelettes as part of the geometric structure, this reaches a critical amount, in which case the protomass becomes a true mass spot. The mass spot can then grow in size, as more energy wavelettes are pressed into the structure.

9. The Mass Spot generally remains a permanent structure. Mass spots will often merge, but not often be destroyed. The mass spots are carried along with the group of energy wavelettes, and therefore are carried with the energy string.

10. Gravity strings emerge from the mass spots. As the energy wavelettes are condensed, the micro gravity strings merge together, creating a single gravity string. This becomes the first true gravity string to emerge from the mass spot into the world beyond.

11. Mass can be converted to energy, using the force of external pressure. This is most commonly from a second magnetic string, which hits the mass spot at just the right angle, and with enough force, to cause the mass spot to break apart. This can also happen within a single string, as the surrounding energy wavelettes add additional pressure, enough to break the mass spot apart.

12. When the mass spot structure is broken apart, the energy compacted within will be released. This will create an explosion. This is the Mass-to-Energy Conversion Explosion.

13. The Mass-to-Energy Conversion can be partial or complete. In a partial conversion, the mass spot can have only part of its mass converted to energy. The amount of energy released will be less, and some of the mass spot will remain in place.
    However, in a full conversion the entire mass spot is converted, and all of the energy is released at the same time. This will result in a powerful explosion of energy.

14. Depending on the amount of mass converted back to energy will result in different energy effects on the surrounding region. Furthermore, the amount of mass remaining will determine how fast it will be to rebuild the mass spot again.

Applications of Mass-Energy Conversions in the Next Chapter

There are several applications of Mass-Energy Conversions, particularly in the areas of Photon Motions. The most important of these applications include:
1. Photon Speed (the Speed of Light)
2. EM Pulsation (the Frequency)
3. Launching of Photon Systems

These Applications of Mass-Energy Conversions will be discussed in the next chapter.

Chapter 5
Applications of Mass-Energy Conversion

Chapter taken from book *Photons in Motion*.

Introduction

Several processes of the physical universe depend on the mass to energy conversions. This means either the conversion of energy into mass, or the conversion of mass into energy.

Some of the most important of these applications are in the arena of photon motions. Several of the primary motions of photons are related to the conversions between mass and energy. More specifically, the mass-energy conversion processes play important roles in three motions of the photon systems:
1. The forward speed of the photon core
2. The pulsation frequency of the field strings.
3. The launching of EM Bursts from an electron.

Each one of these will be discussed briefly below. In addition, some of these will be discussed in greater detail throughout the book.

Speed of Light:
As Pure Energy Inside the Photon Core

Introduction
The "speed of light" is a favorite topic for scientists, and yet nobody has the answer for why the light will travel as fast as it does. Today we can explain the cause of this speed.

The photon system is primarily based on a Photon Core. This Photon Core is a small sphere. This sphere is packed with energy, and sealed completely on the outside. It is because there is so much energy within this small sphere that the photon core can travel at the speed it does.

Photon Structure and Pure Energy
There are two main topics to understand regarding the creation of the photon core: the creation of the photon core structure, and the creation of pure energy within the core. The creation of the photon core is discussed elsewhere. Therefore, in the sections below we will discuss the Pure Energy inside the core.

The photon core is composed of 100% pure energy. There is no mass. This means that there are no mass spots, and there are no protomass pockets. Notice that this pure energy is created from the mass-energy conversion process which was discussed earlier.

In the following sections, the reader will learn the process of creating large amounts of pure energy inside the photon core. Specifically, the process is as follows:

1. <u>Creation of the Photon Core, with Energy Inside</u>

First the photon core is created. A photon core, as with any other particle, is a set of energy strings which loop and become intertwined.

Note that when the photon core is created, it automatically encloses a group of energy strings. This is common during the creation of any particle; it is common for smaller strings to be contained inside the particle as the particle is being created. The difference with the photon core is that the photon core will become sealed.

The fact that the photon core is sealed will create a pressure cooker effect, where all remaining mass is converted to pure energy. The sealed structure will also prevent any energy from leaving the particle. This results in a high energy particle, which will have the same internal energy forever.

2. <u>Same High Energy Within Each Photon Core</u>

These photon cores are created within a star, usually within the highest energy regions. Therefore, the amount of magnetic energy strings per area is extremely high. This means that the amount of energy strings contained within the photon core is extremely high. In other words: the photon core is now packed with energy.

Note also that the number of these energy strings are always the same. This is because of the size of the photon core, compared to the density of the energy strings. The photon core particle is always the same size, and these photon particles are only created in the regions of the star with greatest density of magnetic energy strings.

Thus, we have the same size container, always scooping and enclosing strings in the same regions of energy density. The result is the same number of energy strings scooped up and contained within each photon core.

3. Sealed Photon Core

The core itself is sealed very tightly. The photon core is sealed tightly because it is such a small particle. In all particles, the intertwining strings which lay side by side can merge together. However, there are usually some holes in between the weaving.

The photon core is different. The particle is so small that all of the intertwining particle strings are adjacent. This means that *all* of the looping strings merge together, without any holes between. The result is a solid surface over the entire sphere, like a pinball.

The photon core is therefore sealed very tightly. Because of this, no string ever leaves the core, and no other string enters the core. This ensures that the amount of energy strings in the photon core never changes. This sealed structure also creates the pressure cooker effect which converts remaining mass inside the particle into pure energy.

4. Pressure Cooker Effect

Due to the sealed structure of the photon core, the photon core provides the "pressure cooker effect" which induces the conversion of mass to energy.

The core is small, and tightly sealed. Furthermore, the interior is already packed with energy strings. The tightly sealed particle, with the density of energy strings, will create the pressure needed to convert mass to energy. This is the Pressure Cooker Effect.

5. Explosion of Pure Energy

Remember that every magnetic energy string has some mass spots. Further remember that when enough external pressure is applied to the mass spots, these mass spots will deconstruct, which will then release the energy. There will also be a significant internal explosion.

As illustrated earlier, trillions of energy wavelettes will be released, and all at the same time, from each mass spot. This provides trillions times more energy per each mass spot converted.

This is the case from the pressure of the magnetic strings inside the photon core. These strings press on each other with much force. Each of their mass spots will then explode, with pure energy being released. The net result is that all mass spots in all the energy strings are fully converted into pure energy. This results in a photon core which contains an enormous amount of pure energy inside.

6. Complete Conversion: No Protomass Leftover

We should also point out that there is *no* protomass leftover. Every mass to energy conversion is full and complete. There are no regions of protomass left anywhere.

The process of mass to energy conversion is so complete, for all mass spots, because there is a combination of the sealed photon core with the huge numbers of energy portions. Together these factors create such a strong "pressure cooker effect" that the conversion of mass to energy is certain and complete.

This total conversion is important because without a protomass it is difficult, and likely impossible, to rebuild the mass spot. As discussed earlier, the protomass acts as a seed crystal – we need it in order to encourage other energy portions to join and build the protomass. Without such a protomass, the energy portions are not likely to join together and reconvert into a mass spot.

Thus, because the interior of the photon core has no protomass regions, the interior of the photon core will remain as pure energy.

Therefore: the photon core contains trillions of pure energy wavelettes, pushing on the interior walls of the photon, and driving the photon forward. These energy wavelettes never convert back into mass…and therefore the photon is only pure energy. This is a very high amount of energy, which never converts to mass, and is never let out into space. Therefore, the particle will never slow down.

7. Speed of the Photon

Now we understand why the photon travels at such insanely fast speeds. Much of the energy came from the density of magnetic energy strings. Yet a significant more amount of energy came from the mass-to-energy conversions for all mass spots within all of the strings.

Therefore, the interior of the photon contains only pure energy, no mass. More importantly, the number of energy portions in that one small photon core numbers in the mega trillions. The number of energy wavelettes in that one core is such a high number it is difficult to even fathom. Yet here these mega trillions of energy wavelettes exist, always pushing the photon core forward.

Therefore we have all of this immense amount of pure energy contained in this tiny particle. There are trillions upon trillions of energy wavelettes, all pushing the interior of the walls of the photon core.

This is what drives the photon core at such a fast speed. This is the true physical cause for the famous "speed of light".

Mass-Energy Conversions in Frequency
(The Field Pulsation Process)

Introduction

The other main motion of electromagnetic energy is the pulsation of the field strings. This pulsation process is what creates the frequency. More specifically, the rate of field pulsation is what we measure as the EM frequency.

We will explain the pulsation process in great detail in other publications. At this time it is important to note that the conversions between mass and energy play important roles in the process. Note that both Mass to Energy Conversions and Energy to Mass Conversions are involved in the Pulsation Process.

Photon Core vs Field Strings

As we describe the pulsation process, remember that the Photon System has two components: the Photon Core and the Field Strings. The Photon Core provides the forward motion and speed. The Field Strings create the Frequency.

We described the Photon Core Speed above. Now we will focus on the Field Strings. These field strings exist outside the photon, as passengers on the core. These field strings also create the frequency. They have nothing to do with the speed of the photon.

Basic Pulsation Process

In the EM Pulsation Process there is a repeated expansion and contraction of the field strings. The energy strings first explode outward, due to mass-energy conversion. This is the expansion. Then the energy strings are pulled inward, due to mutual gravitational pull. This is the contraction.

This expansion and contraction will repeat again. This is because the mass spot is being rebuilt, which will allow for a repeated explosion. This repeated action creates a pulsation.

Frequency of EM is Rate of Pulsation

Notice that the overall rate of this pulsation process is the Frequency of Electromagnetic Energy. This pulsation frequency varies from one photon system to another, and the specific cause of that variance is due to the strength of the EM field strings.

Advanced Steps: Rebuilding Mass and the Reversal of Flow

There are a few advanced steps in the process to be aware of now. These are the rebuilding of mass, and the change in direction of flow. Each occurs for different reasons, but both are important.

1. The rebuilding of mass allows a second explosion to occur. Therefore we have mass to energy conversion, then energy to mass conversion, then mass to energy conversion again. This allows us to continue to have new mass-to-energy explosions, and repeat the pulsation.

More specifically: At the same time as the contraction process is occurring, the partial mass left over from the original explosion is quickly being rebuilt. (This is an Energy to Mass Conversion, with assistance from the Protomass). When the energy strings do collide again, there will be another mass to energy conversion, and thus another explosion.

2. The change in flow of the wavelettes occurs because of the influence of gravity strings during the contraction process. The gravity string in this situation cause the wavelettes to reverse direction flow. Then, after the next collision and explosion, the momentum of this reversal of flow will then propel the wavelettes into that direction. This is what results in an expansion in the "opposite direction" from the direction of the previous explosion.

Therefore, each time the magnetic strings collide, the mass spot explodes, which creates an expansion outward. Gravity strings pull the magnetic strings together, and reverse the flow of the magnetic wavelettes. The mass spot is also being rebuilt. Then, on the next collision of magnetic strings, another mass spot explosion occurs, and the magnetic wavelettes are expanded outward again; though this time in the opposite direction as the previous explosion.

This process repeats without change. Thus the pulsation cycle continues.

Further Details on Frequency in *Photons in Motion*

Everything you need to know about Pulsation Frequency is explained (and illustrated) in the book *Photons in Motion*. All details, of every step, in the Pulsation Process are explained there in meticulous detail.

The next few pages will provide a step by step explanation of the EM Pulsation Process.

Pulsation Process:
Focusing on Mass-Energy Conversions

Overview
   In the previous section we discussed the basic process of the EM Pulsation, with emphasis on the conversions between mass and energy. On the following pages, we will expand the details of the process.
   However, there are multiple steps, and many physical factors to look at as the Pulsation Process operates. Therefore, in the following version of the story of EM Pulsation we will focus mostly on the conversions between mass and energy. Other versions with other focus points will be given throughout the book *Photons in Motion*.

1. The Pulsation Strings: Properties
   We begin with the pulsation strings in their normal state. These pulsation strings are electromagnetic energy strings. Specifically, there is one set of electrical energy strings, and one set of magnetic energy strings. Each energy string looks like the diagrams presented earlier.

2. Pulsation Strings: Attached to Exterior of Core
   Each pulsation string is physically attached to the photon core at all times. This is very important to remember. It is this connection which allows the pulsation strings travel with the core, while the photon core is traveling at such high speeds.
   Think of this as a man with a dozen balloons. He is not touching the balloons directly; rather he holds the strings which connect him to each balloon. Similarly, in our photon system we have a photon core and dozens of gravity strings. Each gravity string then connects the photon core to the magnetic energy strings.
   Therefore, the pulsation strings can do their pulsation process (as described below and in other sections) and still remain physically attached to the photon core.

3. Gradual Outward Movement
   We begin with all the energy strings bunched close together. At this stage all energy wavelettes are moving upward and slightly outward. This causes each magnetic energy string to move in a generally upward direction.

4. Mass to Energy Conversion and Explosion

However, the natural migration of energy strings is not enough by itself to create the full outward motion. The main thrust comes from the internal explosion created by mass to energy conversions.

Some of the mass spots of the energy strings will undergo the conversion process of mass to energy. Note that it will only take one mass spot in each energy string to accelerate the string outward at high velocities. In fact, it will only take a few partial conversions, in a few mass spots, to release enough energy for the outward propulsion.

This is the outward explosion (as illustrated earlier) which begins the true pulsation process.

5. Partial Explosions: Protomass Left and possibly more

The internal explosions of the pulsation strings are only partial explosions. This means that not all of the mass spot is converted into energy portions. In fact, not very much conversion is required for the energy string to propel forward much faster than its normal speed.

What remains is a protomass, and in fact a fairly large protomass. The existence of the protomass is essential for rebuilding the true mass spot. Also, because only a small percentage of the mass was converted to energy, there is a large protomass remaining. This will become extremely valuable when we rebuild the mass spot later.

Thus, to review: we have a partial mass to energy conversion. We don't require very much, because a small amount will still produce a significant increase in energy wavelettes. This release of energy wavelettes will propel the energy string outward at a decently high velocity (thus creating the "explosion"), and yet much of the protomass will be remaining.

6. Reconversion of Energy to Mass

The mass spots are only partially converted in the previous explosion. There is a significant amount of each mass spot left. This means that the energy wavelettes will re-convert into mass at a quick rate.

This reconversion process begins almost as soon as the explosion occurs. This is due to the large protomass. Because there is a large protomass, it is easier to rebuild the full mass spot. Many of the nearby energy wavelettes will join the protomass, thereby gradually rebuilding the mass spot to its original size.

7. Gravity String Contraction

The energy strings, having exploded outward, would continue to travel outward on their own forever…except for the many gravity strings. These gravity strings act to pull each of the energy strings back inward together, thus creating the "contraction" process.

Note that there are two sets of gravity strings: those gravity strings holding the magnetic strings to the photon core; and those gravity strings which connect each magnetic string together. Both sets of gravity strings are involved in the contraction. However, the primary set of gravity strings involved are those which connect the neighboring magnetic strings.

Specifically, each energy string has numerous gravity strings extending outward. (This was illustrated earlier). These gravity strings find each other, intertwine, and become gravitationally connected. At this point each energy string is pulling itself toward the other, along this mutual direction of gravitational pull. These two energy strings will eventually pull each other close enough to physically touch. (See earlier chapter).

In addition, as various energy strings start coming closer together, there are other gravity strings waving around which can easily find each other (due to closer proximity). This will increase the number of intertwined strings, and increase the rate at which any two or more energy strings will pull together. (This is a case of Gravitational Acceleration).

Notice that due to the number of gravity strings involved, these energy strings are pulling themselves into each other at a very fast rate. This rate is significant for a later step: the secondary outward explosion.

7b. Reviewing the Gravity String Contraction

To review this contraction step: the entire contraction process is based upon the gravity strings (of each magnetic string) finding and intertwining with other gravity strings (of the other magnetic strings). This gravitational connection begins the mutual gravitational pull among the magnetic strings, where the magnetic strings are pulled inward toward each other. This is the "contraction" process.

In addition, there is a final rate at the moment when the magnetic energy strings collide. This high gravitational pull speed at the time of collision will be an important factor for the next internal explosion.

8. Changing Direction of Energy Flow in Strings

The gravity strings will not only perform the contraction process, they will also reverse the flow of the magnetic energy wavelettes.

As discussed earlier, gravity strings can induce the nearby energy wavelettes to change direction. This is because where the gravity strings pull, the mass spots are pulled. The surrounding energy wavelettes will be pulled as well. Therefore, depending on the direction of the mutual gravitational pull between objects (such as magnetic strings), the surrounding energy wavelettes can be induced to change direction accordingly.

This is the case for all of the magnetic and electric energy strings in the photon system. When these energy strings are pulled together by their gravity strings, many of the individual energy wavelettes will be induced to change direction.

In practical terms, this means: each of the magnetic and electric energy strings will have reversed flow direction by the time they fully contract into collision. This fact is important, as it will propel the energy strings into the opposite direction after the next explosion.

9. Collision of Energy Strings and Next Explosion

The gravity strings will eventually bring together all of the magnetic energy strings. The magnetic energy strings will be contracted until these strings finally collide.

In most cases involving field strings, the field strings will merge together. However, that is not what happens with the EM Pulsation Strings. In this case, the EM Pulsation Strings will explode outward again.

This is because of the angle at which these strings are brought together. They are brought at a slanted angle. Therefore the magnetic strings will not merge directly as most field strings would.

Furthermore, and this is important, the angle at which the magnetic strings come together will hit the mass spots directly. This is similar to a direct missile strike on the mass spot, and the force will cause some of that mass spot to burst open. This creates a partial mass to energy conversion, and therefore a release of energy.

This release of energy from the mass spot, this explosion, will then cause the magnetic strings to be propelled apart again.

Also note that this explosion occurs above the surface of the photon core. Yet at the same time, these field strings are permanently held to the core via gravity strings.

10. Explosion in Opposite Direction

The secondary explosion always occurs in the opposite direction of the previous explosion. The reason for this is the changing direction of the energy wavelettes prior to explosion.

The direction in which the magnetic strings will be propelled after an explosion will be the same direction which these strings were traveling prior to explosion. It is the momentum of the pulsation strings which furthers their direction.

This direction also happens to be the same direction as the direction of contraction, prior to explosion. As we have discussed above, the gravity strings not only contract the magnetic strings together, but also induce the energy wavelettes to change direction with each contraction.

Therefore: the direction of contraction, the direction of the energy wavelettes, and the direction of explosion are all the same. The gravity strings contract the magnetic pulsation strings, which induces the energy wavelettes to reverse flow, which then is the momentum of direction when the mass-energy conversion explosion occurs.

It is for this reason that the direction of each explosion outward is the opposite direction of the direction of the previous explosion. It is also for this reason why the pulsations always occur in opposite directions.

11. Held onto by the Gravity Strings of Photon Core

It is important to know that the Field Pulsation Strings will always be held to the photon core via the gravity strings. Each of the magnetic and electric pulsation strings are held onto the photon core with at least one gravity string. This will never change. Therefore, the pulsation strings will never go flying off the photon system, regardless of the explosion.

12. Mass Spots Slide, Carrying the Field Strings

The Photon Core will also never let go of its magnetic strings for another reason: the mass spots which hold the field strings will slide along the surface, and therefore follow the field strings movement, while holding onto them.

Remember that the mass spots of the photon core structure can slide in any direction. Let us return to our analogy of the man carrying balloons. Imagine our man on an outdoor skating rink. He can skate anywhere around the ice, while still carrying his balloons. If there is a strong wind which blows his balloons in one direction, the man can simply follow the balloons on his skates. Therefore it does not matter where his balloons go, the man can slide anywhere and still hold onto them.

This is exactly what happens with our mass spots, gravity strings, and field strings. Therefore: floating above the photon core are the magnetic strings. These field strings are contracted above the photon core, then exploded outward to the opposite side of the photon core. Yet these field strings will continue to be attached to the photon core. This is because the mass spots on the core surface will simply slide across the surface, to the other side, while continuing to hold onto the field strings. Our balloon man (mass spots) will not let go of any of his balloons (field strings), no matter how fast they move to a new location.

The mass spots simply slide from one pole of the photon core to the other, with each pulsation explosion. The field strings will continue to pulsate, in that other direction, while the mass spots of the surface have migrated with the field strings.

13. Process Repeats; Thus the Pulsation Process

The above steps are the main sequence of steps in the EM Pulsation Process. These steps repeat…again and again and again.

For example: the magnetic strings explode outward, then are contracted inward, then are exploded outward again in the opposite direction. The steps repeat, but directions are reversed, and the magnetic strings are then propelled outward in the original direction. And so the process continues.

This is the Pulsation of the Magnetic Field Strings, which lasts forever. Only when the photon is absorbed and the field strings are plucked off is when this pulsation process stops.

14. Perpendicular Pulsations for Electric and Magnetic Fields

The Pulsation of the Electric Field Strings is exactly the same, just in a perpendicular direction. For example, if the magnetic field strings pulse up and down, then the electric strings pulsate left and right.

Note that these explosions can cross each other quite easily, without interfering with the other. This is because the explosions occur above the surface. Furthermore, the different sizes of the energy wavelettes in each type of field will cause the explosion at different aerial levels above the photon core. Therefore the perpendicular paths will not be on the same aerial level. They will not interfere.

15. Additional Details and Drawings in Later Chapters

Note that the EM Pulsation Process will be repeated several times, in different ways, throughout the book. There will also be additional drawings in later chapters. The reader will come to understand every physical concept involved in the EM Pulsation Process.

Mass-Energy Conversion
for Launching the EM Burst from Electron

Introduction

Another application of the mass to energy conversion is in the launching of the photon systems.

The photon resides inside the electron or proton while the field strings are being attached. During this time, the photon is usually held in place with gravity strings. Therefore in order to launch the photon system, enough energy must be added to break from the gravity strings, and let the photon system fly off into space. There are several mechanisms for this required energy to be added. The mass to energy conversion is one such mechanism.

Holding onto the Photon Core

The photon system can be considered as a high speed train, where the photon core is the train itself and the field strings are the passengers. Continuing with this analogy, the electron and proton are the stations where the photon cores arrive and depart.

However, we know that the photon core always contains immense energy. This is the cause of the fast speed of the photon. Therefore, the photon core must be held in place while the field strings are exiting and new strings are added. This photon core is usually held in place with a webbing of gravity strings. This netting of gravity strings will keep the photon in place, until it is time to launch again.

Field String Assembly and Resulting Frequency

During this time, the field strings of the photon system will be plucked off, and new field strings will be added. Note that as long as the photon core remains essentially in place, the field strings will come and go. It will only be when that photon system is launched again that those field strings will be the fixed passengers. These field strings, once launched, will determine the frequency of pulsation.

Launching the Photon System With Enough Magnetic Field

There are two main ways to apply enough energy to launch the photon from its gravitational webbing. The first is having a strong enough magnetic field. The second is mass to energy conversion.

In the first case, amount of field strings attached to the photon core is enough to launch the system. The total of the field strings will be enough to pull the photon system (photon core and the field strings) away from the gravity strings in the electron. The photon system is then launched.

Mass to Energy Conversion in Launching Electron

The second method is for a mass to energy conversion to occur. In this process, one of nearby magnetic strings has a mass-to-energy conversion. The energy released from this conversion is enough to break the gravitational webbing which holds the photon, which allows the photon system to then launch.

We will begin discussing this process by looking at the nearby free floating magnetic strings. There are many free floating magnetic strings inside the electron. These free floating magnetic strings provide the internal energy of the electron, and push on the walls of the electron to create the speed. These are the magnetic strings which can undergo mass to energy conversions, which then acts as launching fuel for the photon system.

We can think of this process as lighting a fire under the gravity strings. Imagine a balloon held by several ropes. When we set fire to the ropes, the ropes disappear, and then the balloon rises. A similar operation occurs for the mass-energy conversion and the photon launch. A mass to energy conversion from the free magnetic strings will release enough energy to break apart the webbing of gravity strings. This allows the photon system to break free, and fly out of the electron into space.

More Detailed Understanding of the Process

We will discuss this process in greater detail at this time. Any two of the free floating magnetic strings can collide in such a way as to induce a partial mass to energy conversion. This will release additional energy wavelettes into the nearby region. If this occurs next to the webbing of gravity strings, then the gravity strings will be blasted apart. The photon system will then be free to launch. Within nothing holding the photon core back, this photon will quickly fly out of the electron.

The added wavelettes will also give the photon a pointed direction, by energy by pushing the photon up and out a particular hole.

Mass-to-Energy Conversions and Photon Launch Review

Therefore: the mass to energy conversions can play an important role in the launching of the photon system.

All photon cores are held in place by gravity strings inside the electrons and protons. Any two magnetic strings near the photon can collide in such a way as to induce a partial conversions of mass spots to energy. The released wavelettes from the mass-energy conversion can then break apart the gravity strings. The separated gravity strings no longer holds the photon system in place, which then allows the photon system to launch based on its own internal energy.

Chapter 6
Gravity Strings in Mass-Energy Conversions

Chapter taken from book *Photons in Motion*.

Overview

Gravity Strings always emerge from Mass Spots. Therefore, when the mass spots are deconstructed the gravity strings may also change. In this chapter we will look more closely at some of the possibilities.

Note that the mass to energy conversions involve primarily the magnetic energy or the electric energy. These are the energies compacted into the mass spot. However the gravity strings will also merge, and become a single longer string. Therefore everything is related.

Energy to Mass Conversion and Resulting Gravity Strings

Introduction

We will first look at the Gravity Strings when the Mass Spot is being created. Remember that the magnetic energy wavelettes have micro gravity strings connecting them loosely together. Therefore, when the magnetic wavelettes are condensed together, the gravity strings are also condensed. However the structure is different. While the magnetic energy changes structure into a mass spot, the gravitational energy becomes a longer string.

Process of Creating Mass and Longer Gravity Strings

More specifically: the magnetic energy wavelettes are pressed from the outside. This causes the wavelettes to merge and change form, into the structure of the mass spot. Meanwhile, the microgravity strings find each other, and merge taller. This becomes a single, taller, gravity string.

Gravity String is Firmly Embedded Within the Mass

The bottom of this gravity string is firmly embedded in the center of the mass spot. This is where the first magnetic energy has been constructed into mass, and therefore the first micro gravity string are there. As more energy converts to mass, this will add layers to the sphere, and add length to our magnetic string. Therefore the bottom of our gravity string is strongly embedded within the mass spot, while the top extends above.

The length of this gravity string will depend of course on the amount of the magnetic energy condensed into mass.

Mass-to-Energy Conversion and Resulting Gravity Strings

Introduction

Next we will look at the conversion of Mass to Energy, and what will happen to the existing Gravity Strings.

Remember that a gravity string must be physically attached to a mass spot. Thus, if a portion of a mass spot is converted to energy wavelettes, then that mass portion is essentially "taken away". What is the gravity string to do? The first thing to consider is whether the mass to energy conversion is a partial conversion, or a complete conversion.

1. In a partial mass-to-energy conversion, the gravity string remains intact. Though some of the mass is taken away, the gravity string remains. The gravity string can continue to perform its functions.

2. However, in complete mass-to-energy conversion the gravity string will be ejected. This is similar to a hair which has been separated from its skin. Note that this free gravity string will usually merge with other gravity strings soon after being ejected.

3. Also in the complete conversion, the gravity string (prior to ejection) may be attached to another in the "Y" shape, and will therefore dangle awhile. Eventfully this dangling end will attach itself to another mass, or merge with another gravity string.

Each of these situations will be described below.

1. Partial Mass to Energy Conversion: Gravity String Intact

If it is a *partial* mass to energy conversion, then some of the mass spot will remain. In this case, the gravity string will remain intact, and will remain as the same length. Although much of the original mass has been converted back into to energy, the gravity string remains as it is.

Notice that Gravity Strings will remain, even when some of the mass is deconstructed. This is similar to a tree on an island. The water can take out much of the land, but the tree in the center still remains. The gravity string in this situation is similar. Much of the surrounding mass spot can be converted, while the single gravity string in the center remains untouched.

The reason for this stability of the gravity string is that the bottom end of the gravity string is firmly embedded within the center of the mass spot. (See the formation process above). Therefore the outer portions of the mass can be reconverted into energy, without touching the gravity string at the center.

Furthermore, the gravity string is no longer based on the magnetic energy wavelettes. Although the magnetic energy did initially bring the micro gravity strings, at this time the magnetic wavelettes have given up their gravity to the gravity string in the center. The magnetic wavelettes are no longer connected to the gravity string. We can therefore take away much of the magnetic energy around the outer portion of the mass spot, and the gravity string in the center will remain as it is.

2. Complete Mass-Energy Conversion: Gravity Strings Ejected

Introduction

In a *complete* Mass-to-Energy Conversion, the Gravity String will be ejected. However, what happens next with the gravity string can be several possibilities.

Notice that when there is a *complete* conversion of mass spot into energy, then none of the mass will remain. All of the mass will have been converted to energy. This will eject the gravity string outward as an intact gravity string.

Once ejected, the gravity string will likely merge with nearby gravity strings. In rare cases, the ejected string becomes a free flowing gravity string.

Ejecting the Gravity String

As described above: when a complete mass to energy conversion takes place, all of the mass is reconverted back into energy wavelettes. This process removes any stable location for the gravity string to be connected. This also creates enough energy to push the gravity string away. Therefore, the Gravity String is ejected. It remains as its complete length of gravity string, yet is not attached to its mass location.

There are several options for what happens next: 1) The gravity string will merge with adjacent gravity strings; 2) the gravity string was connected to "Y", and will dangle for a while; or 3) the gravity string will become a completely free gravity string.

1. Complete Merging of Ejected Gravity String with Adjacent

The most common option is that the gravity string will merge with other nearby gravity strings. It is most likely that the ejected gravity string will quickly bump into an adjacent gravity string. These strings will merge completely. Both strings merge together completely; and the total length of this gravity string at this location will be a correspondingly greater length.

2. Dangling of the "Y" Leg, then Complete Merging

The second likely option is that this gravity string was part of the inverted "Y", and will now dangle in the air for a while. This option requires a bit more visualization, but the concept is rather simple.

Remember that when two gravity strings merge, this creates an inverted "Y" shape. At the level of mass spots, this means each of those legs is attached directly to a mass spot. Therefore, when one mass spot disappears, then this leg of gravity string will have no place to stick. It will then dangle in space for a while.

However, this dangling will not occur for long. This dangling leg will then merge with adjacent gravity strings, as in #1 above. Thus, instead of just a "Y" shape…the two gravity strings merge completely together.

In the analogy of the car lanes, this would be letting ALL cars from the right lane to merge into the left lane. At the merging point, one car pauses and lets the entire set of other cars join his lane ahead of him. This is what the gravity strings will do. The ejected leg of the Y will continue to flow upwards, and into the "lane" of the other gravity string.

Notice that the entire length of the gravity string merges at this time, which is different from the case when there was just the Y shape. The result is a single, and much longer, gravity string. (This gravity string is extending from the second mass spot).

3. The Free Gravity String

The third option is for the ejected gravity string to become a completely free gravity string. If the gravity string is ejected into space, and does not encounter any adjacent gravity strings to merge with, then the gravity string will become a truly free gravity string.

This gravity string will then float around freely through the air. These gravity strings can also form loops and tiny balls of gravitational energy.

How long will these free gravity strings remains free will depend on the environment. In most regions there are gravity strings everywhere. Therefore the free gravity string will likely bump into adjacent gravity strings somewhere, and merge completely with those strings. It is for this reason that free gravity strings are rarely found.

However, free gravity strings and tiny gravity balls can exist. They are extremely rare but they can be found, particularly in deep space.

Chapter 7
Review and Conclusion

Overview: A New Physical Understanding of Energy and Mass

In this book we provided a new understanding of Energy and Mass. or the first time, we have physical understanding of these concepts. What was once presented as abstract or purely mathematical, we have given concrete form and substance.

For the first time, we can visualize energy and mass. We can see these energies in operation. We understand flow, merging, and conversions.

Furthermore, the descriptions in this book are the most accurate at this time. Never has energy and mass been understood with such accuracy, and in such very concrete physical ways.

The readers will also notice that here is no need for the complex dimensions or other overly complex ways of understanding energy. The information in this book is the best, the most accurate, descriptions of energy and mass available.

Having understood these physical details, the reader can proceed to develop even deeper understandings of science; as well as develop more highly advanced technologies.

Let us now review the main concepts of Energy, Mass, and the Conversions which we discussed in this book.

Universal Energy Creates All Things

We begin with the Universal Energy. This Universal Energy is the basis for all things. All entities and motions can be create from this Universal Energy. Each type of energy, each field, each particle, and each motion, will all come from this Universal Energy.

The nature of this Universal Energy will be disclose in a future book. For the moment, know that this Universal Energy does exist.

The Universal Energy is the basis for each type of energy. Each of types of energy that we know of is a specific amount of the Universal Energy, as well as the geometric shape. These energy types will then become: energy fields; particle structures; mass spots; internal motions; and bonds between particles.

Energy Wavelettes

Energy exists in the form of wavelettes. We can call these Energy Wavelettes, Energy Units, or Energy Fish. They can be considered as Energy Fish because each of these wavelettes vibrates and moves forward similar to a fish. Furthermore, similar energy wavelettes will travel in groups, similar to fish in the ocean. These groups of energy fish are what we know of as Energy Strings.

Difference in Types of Energy Wavelettes

There are three main type of energy: Gravitational, Electric, and Magnetic. Note that the nuclear energy is created from the magnetic energy. Also note that the mass spot is created from either the electric or the magnetic energy.

Each of these types of energy are composed of the same primary material: the Universal Energy. The difference between the types of energy is therefore the amount of this Universal Energy, as well as the geometric structure of the wavelette.

When an individual wavelette has more of the Universal Energy, then it will be larger. It will also be a "stronger" form of energy. Conversely, when an individual wavelette has less of the Universal Energy, then it will be smaller, as well as being a "weaker" form of energy. This provides the basic differences in the types of energy.

Therefore we have the following characteristics for each type of energy:

1. Gravitational Energy: small amount of the Universal Energy; very small energy wavelette; simple shape; weakest of all known energies.

2. Electrical Energy: medium amount of the Universal Energy; medium sized energy wavelette; slightly complex shape; medium-high strength energy wavelette.

3. Magnetic Energy: high amount of the Universal Energy; large size energy wavelette; complex shape; very high strength energy wavelette.

4. Nuclear Energy: very high amount of Universal Energy; very large size energy wavelettes; very complex shape; highest strength wavelette. Note that nuclear energy is actually a compacted form of magnetic energy, and is only created under high pressure conditions.

5. Mass Spots: extremely high amount of Universal Shape; much larger entity than the energy types. Inner-locking structure is highly complex, yet the overall structure creates a sphere. Extremely high strength, but only noticed when energy is released.

Note that the mass spot is technically not an Energy String, but is similar, and listing in this way the reader can better understand the progression.

Converting from One Type of Energy to Another

Because each type of energy is created from the Universal Energy, any one type of energy can be converted into another. Smaller wavelettes can be combined into a larger wavelette. Conversely, larger wavelettes can be broken apart into smaller wavelettes.

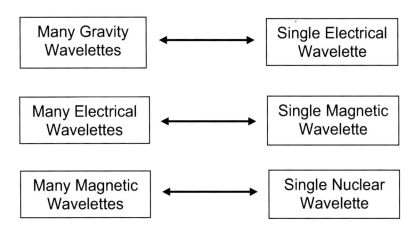

Mass Spots

Mass is energy which has been concentrated into complex and tightly interconnected structures. These regions of concentrated energy are the areas of mass, known as the Mass Spots. Mass spots are usually spherical, but can be other shapes. They will often be viscous.

Mass spots are created from the wavelettes of various types of energies: Electrical, Magnetic, and Nuclear. Note that gravitational energy does not have enough strength to create a mass spot.

Mass spots are created by applying external pressure of outer wavelettes on the inner wavelettes. These inner wavelettes are then condensed and restructured into the new structure of mass.

Energy Strings: New Concepts of Structure

The True Structure of the Energy String is very much different from what is taught in the current physics. The true Energy String structure needs only 3 dimensions, and is very Newtonian. The True Energy String is able to create all known particles, fields, and processes in the universe.

This True Structure of the Energy String is composed of energy wavelettes. These energy wavelettes, also known as Energy Fish, will group together and travel together. A grouping of these energy wavelettes therefore becomes known as an "Energy String".

The types of Energy Strings depend on the type of Energy Wavelettes which are grouped together. This creates the Energy Strings of Gravity Strings, Electrical Strings, Magnetic Strings, and Nuclear Strings.

These Energy Strings move about on their own, as would a group of fish in the ocean. However, their paths can adjusted by the influence of other Energy Strings.

Note that Mass Spots always exist within these Energy Strings. This is an important structural concept of the True Energy String. The percentage is approximately 80% energy wavelettes to 20% mass spots, per area, in a magnetic energy string.

Furthermore, the mass spots will be moved along with the wavelettes in the energy string. The mass spots are pushed along by the surrounding wavelettes. Therefore, where the Energy Fish go, the Mass Spots will also be pushed along.

Energy Strings: Merging and Separating

The Energy String has no physical borders. Rather, the "string" is group of individual wavelettes. Therefore groups of wavelettes can merge together. This is how the Energy Strings merge together.

Similarly, an Energy String can easily break apart. The Energy String is just a group of wavelettes, therefore any smaller group of wavelettes can break off from the main group at any time. This is usually induced by nearby energy strings. This is how the Energy Strings will break apart into smaller segments.

Most energy Strings can grow either in thickness or length. Gravity strings will only grow in length. Nuclear energy strings tend to grow in thickness. Energy strings can be much smaller than a neutrino, or the size of stars. Gravity strings can be long enough connect stars and galaxies.

Energy Strings: Flow

The direction of flow for the Energy String is the direction which the majority of energy wavelettes are traveling at that time. Most often the individual wavelettes are traveling in the same direction, however some wavelettes can be pivoting or moving away in other directions. This will affect various properties of the energy string.

Energy Strings: Applications

The Energy Strings can be used for various applications. These include:
1. Particle Structure
2. Internal Energy
3. Attached Energy Fields
4. Gravitational and Magnetic Pulls

1. Particle Structure

The Particle Structure is created by intertwining energy strings. The magnetic energy strings and electrical energy strings will intertwined and loop together, which creates the particle structure.

2. Internal Energy and Driver Strings

Internal Energy for a particle is the magnetic energy strings which are contained within the particle. These are also the Driver Strings, as these strings push on the interior of the particle walls and propel the particle forward.

3. Attached Energy Fields

The Energy Fields are actually Energy Strings. Any one grouping of Energy Wavelettes can be called an Energy String or an Energy Field. However, the term "field" usually applies to an Energy String only when that string is attached to a particle. Free floating strings can be given other names. There are three types of Energy Fields: Magnetic, Electric, and Gravitational. Each are groups of energy strings which are attached to the particle in different ways. (Details will be shown in other publications).

The mathematical sign given to a field simply indicates the direction which the energy string is flowing. For example, a set of Electrical Energy Strings flowing out of a particle is "negative electrical energy"; while the same type of Electrical Energy Strings flowing into a particle is "positive electrical energy".

4. Gravity Strings

Gravity Strings are a special case of Energy Fields. All of Gravitational Energy is contained within these Gravitational Energy Strings.

Gravity Strings are attached to the Mass Spots. More specifically, the bottom of the gravity strings is embedded within the mass spot, and extends into the world. From there, adjacent Gravity Strings will merge, becoming longer and longer. These Gravity Strings can then reach distances of many miles.

See the book "Introduction to Gravity Strings" for complete details

Gravity Strings and Magnetic Energy Strings

Gravity Strings and Magnetic Energy Strings have a special type of physical structure. This structures is similar to a tree with branches. Specifically: the Magnetic Energy is the main tree, with thinner Gravity Strings which extending outward.

Remember that all Magnetic Strings have mass spots, and all mass spots have Gravity Strings. Also remember the gravitational energy wavelettes are much smaller and weaker than magnetic energy wavelettes. Therefore: the Magnetic Energy String is very much larger than the Gravitational Energy String. The structure is of a Magnetic String Tree with several Gravitational String Branches.

A similar structural relationship exists with the Electrical Energy Strings and Gravitational Energy Strings.

Converting Energy to Mass

Energy is converted into mass by applying external pressure of outer wavelettes on the inner wavelettes. These inner wavelettes are then condensed and restructured into the new structure of mass.

The amount of energy compacted into mass can vary, depending on the available energy wavelettes and the amount of external pressure. This also means the resulting size of the mass spot can vary. Larger mass spots will have more energy compacted. However the amount of energy compacted into any mass spot is quite enormous.

Converting Mass to Energy

Mass spots can be converted to energy again. This requires the right amount of external pressure applied at the exact locations.

Most of the time these mass spots are quite stable. The mass spots exists, with their extending gravity strings, for decades and centuries. However, given the right circumstances these mass spots will be broken apart, which will unleash great amounts of energy.

The process involves the application of external energy. When enough additional external energy is applied, directly to the mass spot, then the mass spot will deconstruct and implode. The wavelettes within the mass will burst forth like a volcano.

Notice that external pressure is needed to both create and destroy the mass spots. However, there are some differences. For example, the mass spots are usually created within a group of wavelettes (within one Energy String). Yet the destruction will usually require a second group of wavelettes (second Energy String hitting mass spot of the first String).

Applications of Mass Energy Conversions

There are several natural applications of the Mass-Energy Conversions. The Photon System uses these conversions in several areas. These include:
1. Speed of Light, as Internal Energy of the Photon Core.
2. Launching of the Photon System from Electron
3. Pulsation Process of the EM Field Strings

Unified Field Solution

These physical descriptions of the Energies will of course lead us to a Unified Field Solution. For decades, scientists have been looking for a way to unify all the energies. We can now see that this is easily done. With the descriptions and illustrations of energy provided in this book we can easily create a true and accurate solution to the Unified Field.

We have hinted at this solution to the Unification of Energy throughout this book. The details to this solution have already been worked out, and will presented in a future publication.

Theory of Everything: The Solution

The other great quest of scientists is to develop a "Theory of Everything". In this theory, all physical entities and processes will be explained by only a few concepts. Again, we can see that using the concepts discussed in this book, the Theory of Everything is easily within reach. We have provided hints of this Theory throughout the book.

Indeed, this Theory of Everything has already been fully developed by the Author, and will be presented in its complete form in a future publication. All other books in the Series will be leading to that point.

Published Books in the Physics for the Next Millennium

Readers are encouraged to read all of the books in the series "Physics for the Next Millennium". This Series of books presents completely new discoveries in physics; replacing the outdating teachings with more accurate understanding. Some of the specific books include:

- Introduction to Gravity Strings
- Momentum Understood as Energy Strings
- Understanding Dimensions

Future Books Related to the New Physics

There are also several books of the New Physics which are in progress. Note that the solutions have been developed for each of these topics, it is only the writing and illustrations which need to be completed. These books include:

- New Model of the Atom
- How to Understand Time
- Photons in Motion
- Unified Energy: The Solution
- Theory of Everything: The Solution

Websites

Readers can learn more information about these topics (and many others) at the following locations:

Amazon Author Page
https://www.amazon.com/author/markfennellvisionary

Main Website
https://markfennellvisionary.com

Blog
http://markfennell.blogspot.com

Video Channel: All Thing Energy
http://www.youtube.com/channel/UCk5ckPqF4oD0JoJMSBi2Zcg

Conclusion

Energy is the basis for all things. All of the universe, in all aspects, is energy. Yet for many centuries, the nature of energy has been discussed as abstract rather than real physical entities. Furthermore, traditional scientists have created complicated methods of working with energy, when the reality is much simpler.

Today we finally have the true and accurate solutions to the mysteries of Energy. We now have physical descriptions and illustrations of energy. We can finally understand the structures and processes of energy.

This book is the first publication to offer these solutions and physical descriptions. The implications of what you have read are profound.

Use this information wisely.

Mark Fennell

October 22, 2018

Made in the USA
Columbia, SC
04 March 2021